Vipul Kakkar
General Topology

Also of Interest

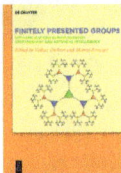

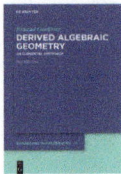

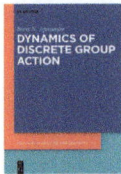

Vipul Kakkar

General Topology

Metric Spaces, Dedekind Cut, Isometries, and Hilbert Cube

DE GRUYTER

Mathematics Subject Classification 2020
54-01, 54B05, 54B10, 54E45

Author
Dr. Vipul Kakkar
Department of Mathematics
Central University of Rajasthan
305817 Kishangarh
Rajasthan India
vplkakkar@gmail.com

ISBN 978-3-11-163606-1
e-ISBN (PDF) 978-3-11-163608-5
e-ISBN (EPUB) 978-3-11-163619-1

Library of Congress Control Number: 2024950869

Bibliographic information published by the Deutsche Nationalbibliothek
The Deutsche Nationalbibliothek lists this publication in the Deutsche Nationalbibliografie;
detailed bibliographic data are available on the Internet at http://dnb.dnb.de.

www.degruyter.com
Questions about General Product Safety Regulation:
productsafety@degruyterbrill.com

To my son Vanjul

Preface

Although the topology has originated in different manners, it is a good idea to learn metric space first and then topology. Therefore the present book is completely dedicated to metric spaces and their topology. This subject is a necessity for many mathematical specialties and other fields. One is likely to approach the topic from their own perspective.

The present book is an outcome of my lecture notes, which were prepared to teach topology at the Central University of Rajasthan, India. This book tries to answer the questions asked during the classroom teaching, either by me or by the students.

The book starts with a basic fundamental question: "What is a set?" Why an axiomatic system is required. In the first chapter, the Zermelo–Freenkel axiomatic system is discussed. The complete description of the various number systems has been discussed in the first chapter. The constructions of the real number system with the help of Dedekind cuts and Cauchy sequences are discussed in the Chapters 1 and 5, respectively.

Chapter 2 starts with the definition of a metric space. In this chapter, various examples and properties of the metric spaces and normed spaces are discussed. Also, the distances between the sets in metric spaces and the Gromov–Hausdorff distance are discussed.

Chapter 3 is devoted to maps between metric spaces. The first section discusses continuous maps and their properties. The second section discusses homeomorphisms and their properties. The third section discusses equivalent metrics. The notions of Lipschitz equivalent and topological equivalent metrics are discussed. In the fourth section the isometry between metric spaces is discussed. The fifth section discusses the finite metric spaces and their embedding in Euclidean spaces.

In Chapter 4, metrics on the product and quotient sets are discussed. Again, the metric-preserving maps are discussed in this setting.

Chapter 5 is devoted to sequences in metric spaces. The first section deals with convergent sequences and their properties. The second section discusses complete metric spaces and their properties. In the fifth section the completion of a metric space is discussed. In the fourth section the real number system is obtained via the completion of the rationals, which is an incomplete metric space. In the fifth section, the Baire category theorem and its applications are discussed.

Chapter 6 is devoted to compact metric spaces. The first section discusses their properties. In the second section, some equivalence of compactness is discussed. In the third section the Hilbert cube is discussed, and it is proved that a compact metric space can be embedded in the Hilbert cube as a closed set. The fourth section discusses the Cantor set, and it is shown that a compact metric space is a continuous image of the Cantor set.

The last chapter is about connected metric spaces. Various properties of connected spaces are discussed in the first and second sections. The last section discusses the path-connected metric spaces.

A book cannot be completed without the help of others. First of all, I am thankful to my teachers, Prof. Ramji Lal (retired) and Prof. R. P. Shukla, University of Allahabad,

https://doi.org/10.1515/9783111636085-201

Prayagraj, for their constant encouragement. I am also grateful to Prof. D. P. Choudhary (retired), University of Allahabad, Prayagraj, for pointing out various errors in this book. However, if there are any errors left, I am the only one responsible for them. I will be thankful to the readers if they can let me know any kind of errors in the book. I am very thankful to the editorial team of De Gruyter, especially Dr. Ranis N. Ibragimov. He was very helpful at various stages. I am thankful to my parents, Mr. V. K. Khattri and Mrs. Gaytri Khattri, for their constant moral support. Last but not least, my wife Vinny has my sincere gratitude for her patience, unwavering support, and encouragement all across my life.

Central University of Rajasthan   Vipul Kakkar
India

Contents

1 Theory of sets

In this chapter, we will very briefly study the axiomatic set theory and construct some important number systems.

1.1 Need of axiomatic system

It is a general curiosity of various people "What do you do with math?" or "What is math?" A sophisticated answer of this question may be "It is the study of structures." A structure is a system of some kind of objects that have certain relationships defined for those objects or their pairs, triples, etc. The groups, rings, metric spaces, or topological spaces are a few examples of mathematical structures. The content of the object that constitutes the structure under study is neither highlighted nor denied during the study of a structure. Although it is not necessary to know what oxygen is throughout life, but it is a good idea to know what you are breathing at least once in your life.

The theory of sets as a formal study of mathematics started with the work of Georg Cantor. It was initiated from the questions of analysis.

Suppose that $\sum_{n=-\infty}^{n=\infty} a_n e^{\mathrm{int}}$ and $\sum_{n=-\infty}^{n=\infty} b_n e^{\mathrm{int}}$ ($0 \leq t < 2\pi$) are the Fourier series such that

$$\sum_{n=-\infty}^{n=\infty} a_n e^{\mathrm{int}} = \sum_{n=-\infty}^{n=\infty} b_n e^{\mathrm{int}}.$$

Is $a_n = b_n$ for all n? Cantor in 1870 showed that this is true. He defined a set as follows:

By a set we shall understand any collection into a whole X of any definite, distinct objects x (which will be called elements of X) of our intuition or of our thought.

We may further ask what is meant by objects, distinct objects, collections, etc. Therefore a definition is required. It seems that there must be a point where we can start. Such concepts are called the primitive concepts. The term "set" will be one of the primitive terms for us. Let us consider one example (nonmathematical) that we can face.

Example 1.1.1. One day a little kid called Mr. Vanjul asked his mother after seeing a baby cat "Who is mother of that baby cat?" His mother replied that the black cat is the mother of that baby cat. Mr. Vanjul further asked "Who is mother of that black cat?" His mother replied that she did not know but it certainly exists. As soon as Mr. Vanul got the existence of a mother, he started asking the existence of mother of mother.

This seems to be an infinite loop of query, but very cleverly his mother made an assumption or an axiom that there is a cat who is not a baby of any cat. In the evidence, she told a story of Adam and Eve and assumed that there must be Adam and Eve like creature for cats also.

https://doi.org/10.1515/9783111636085-001

There are situations in which certain assumption can give rise to an adverse situation. These are called paradoxes. Let us consider the following (nonmathematical) examples.

Barber's paradox
There is a barber who shaves all those and only those who does not shave themselves. This statement can not answer "Who shaves the barber?" If he shaves himself, then he contradicts the statement. If he does not shave himself, then according to the statement, he must shave himself, again a contradiction.

King–thief paradox
One day, a thief was caught and taken to the king's court. The king instructed his minister to inquire with the thief about the truth. He ordered a life sentence for speaking the truth otherwise hanging the thief to death. The thief was brought before the king the following day. The king asked "What would you like to say?" The thief responded "I will be hanged until death, my lord!" Upon hearing this, the king had no choice but to leave the thief, as if his statement were false, then he would not be hanged, but he should have been hanged for the false statement. On the other hand, the king would contradicted himself again if it was true that he would be hanged as for the truth, he should have sentenced him to life in prison.

An axiomatic system for the theory of sets is required to study and develop a theory that is free of paradoxes.

1.2 Logic

The logic may be defined as a language of science. In any language, we usually find some symbols and their combination, which we call a sentence, such that our idea can be expressed and shared. In our particular situation, we are interested in those sentences for which we can affirm whether they are true or false but not both. These sentences are called mathematical or logical statements or propositions. By a statement we will always mean a mathematical statement. We will combine the statements using the connectives. These are "and", "or", "not", "if . . . , then", "if and only if".

If p denotes a statement, then *not p* is called the negation of p. We denote is by $\neg p$. The negation of a true sentence is false, and that of a false sentence is true.

Let p and q be statements. We denote *p and q* by $p \wedge q$ and call the conjunction. The conjunction of two sentences is true if and only if both statements are true. If we assign "T" and "F", respectively, for the true and false statements, then we can represent this in Table 1.1, which we call a truth table:

Table 1.1: Conjunction.

p	q	$p \wedge q$
T	T	T
T	F	F
F	T	F
F	F	F

We denote *p or q* by $p \vee q$ and call the disjunction. The disjunction of two sentences is true if and only if at least one of the statements is true. By $p \Rightarrow q$ we mean that the statement *p* logically implies *q*. This is called the implication. The implication $p \Rightarrow q$ is false if *p* is true and *q* is false; otherwise, it is true. There are different ways of stating this; for example, "*p* implies *q*", or "if *p*, then *q*", or "*p* is a sufficient condition for *q*", or "*q* is a necessary condition for *p*". In $p \Rightarrow q$, *p* is called a hypothesis or antecedent, and *q* is called a conclusion or consequent. The truth table of the implication is shown in Table 1.2:

Table 1.2: Implication.

p	q	$p \Rightarrow q$
T	T	T
T	F	F
F	T	T
F	F	T

From the truth table of implication we can observe that if the hypothesis is false, then the implication is always true irrespective of conclusion. There will be several places where we will be using this fact in proving some statement. A sentence whose truth-value is always true is called as a tautology. A sentence whose truth value is always false is called a contradiction.

Two mathematical statements *p* and *q* are called logically equivalent if $p \Rightarrow q$ and $q \Rightarrow p$. We denote it by $p \Leftrightarrow q$. We can check that the implication $p \Rightarrow q$ is logically equivalent to $\neg p \vee q$. The following laws hold for the connectives conjunction, disjunction, and negation.

(i) *Idempotent law*
    (a) $p \wedge p \Leftrightarrow p$, (b) $p \vee p \Leftrightarrow p$.
(ii) *Associative law*
    (a) $p \wedge (q \wedge r) \Leftrightarrow (p \wedge q) \wedge r$, (b) $p \vee (q \vee r) \Leftrightarrow (p \vee q) \vee r$.
(iii) *Commutative law*
    (a) $p \wedge q \Leftrightarrow q \wedge p$, (b) $p \vee q \Leftrightarrow q \vee p$.

(iv) *Distributive law*
  (a) $p \land (q \lor r) \Leftrightarrow (p \land q) \lor (p \land r)$,
  (b) $p \lor (q \land r) \Leftrightarrow (p \lor q) \land (p \lor r)$.
(v) *Identity law*

$$(a)\ p \land T \Leftrightarrow p, \quad (b)\ p \lor F \Leftrightarrow p,$$
$$(c)\ p \land F \Leftrightarrow F, \quad (d)\ p \lor T \Leftrightarrow T.$$

(vi) *De Morgan's law*
  (a) $\neg(p \land q) \Leftrightarrow \neg p \lor \neg q$,   (b) $\neg(p \lor q) \Leftrightarrow \neg p \land \neg q$.
(vii) *Negation law*
  (a) $p \land \neg p \Leftrightarrow F$,   (b) $p \lor \neg p \Leftrightarrow T$,   (c) $\neg(\neg p) \Leftrightarrow p$.

The following rules of inference are used in deriving the conclusion from a given hypothesis.
(i) *Tautology:* $p \Rightarrow p$.
(ii) *Rule of detachment:* If p is true and $p \Rightarrow q$, then q is true.
(iii) *Rule of syllogism:* If $p \Rightarrow q$ and $q \Rightarrow r$, then $p \Rightarrow r$.
(iv) *Contrapositive rule:* $p \Rightarrow q \Leftrightarrow \neg q \Rightarrow \neg p$.

Let $P(x)$ denote an expression that involves some variable x that becomes a mathematical statement when we substitute a suitable object for x. We will use two logical quantifiers called the universal quantifier $\forall$ and the existential quantifier $\exists$. By $(\exists x, P(x))$ we mean that "there exists an object x such that $P(x)$ is true". By $(\forall x, P(x))$ we mean that "for all objects x, $P(x)$ is true". We can observe that the negation of $(\exists x, P(x))$ is $(\forall x, \neg P(x))$ and the negation of $(\forall x, P(x))$ is $(\exists x, \neg P(x))$.

1.3 The Zermelo–Fraenkel axiomatic system

The works of Georg Cantor and Richard Dedekind paved the way for the modern set theory. In its beginning the paradoxes appeared in the set theory. Therefore set theory required an axiomatic system that was free from paradoxes. In this section, we present one axiomatic system called the Zermelo–Fraenkel axiomatic system. There are also other axiomatic systems for the set theory. The term "set" will be a primitive term for us. Also, the sense of membership will also be primitive for us. By $x \in A$ we will mean that "the object x is a member of A". There are other ways of saying the same thing, for example, "x belongs to A", "x is in A", etc. By $x \notin A$ we mean that "x is not a member of A". We have one more predicate that we take as a primitive term. This predicate has the sense of equality. We denote it by $=$.

Axiom of extension. *If every element of the set A is an element of the set B and every element of the set B is an element of the set A, then the set A is equal to the set B.*

This simply says that a set is completely determined by the expansion or extension of its elements. This axiom indeed relates the predicates $\in$ and $=$.

Axiom of existence. *There is a set with no elements.*

By the axiom of extension we can observe that the set containing no element is unique. This set is called the empty set or null set. We denote it by $\emptyset$.

Let $P(x)$ denote a property that involves some variable x.

Axiom of specification. *Given a set A, there is a set B that consists of all the elements x from the set A satisfying P(x).*

By axiom of specification, $\{x \in A \mid P(x)\}$ is a set. By the axiom of extension we can note that such a set is unique. We must definitely point out that we are not saying that there is a set for any property. Let us check if we have this as an axiom. This implies that $M = \{x \mid x \notin x\}$ is a set. Then either $M \in M$ or $M \notin M$. If $M \in M$, then $M \notin M$, and if $M \notin M$, then by the defining property of the set $M \in M$. Therefore we get a paradox called as Russell's paradox. Indeed, this axiom restricts in defining the set of all sets. On the contrary, if S is the set of all sets, then by the axiom of specification $\{x \in S \mid x \notin x\}$ is a set, but this leads to a contradiction.

If A and B are sets, then by the axiom of specification $\{x \in A \mid x \notin B\}$ is a set. This is denoted by $A \setminus B$ or $A - B$ and called the complement of B in A.

We say that a set A is a subset of a set B if $x \in A$ implies $x \in B$. We denote this by $A \subseteq B$. We can easily note that $A \subseteq A$. By $A \subset B$ or $A \subsetneq B$ we mean $A \subseteq B$ and $A \neq B$. If $A \subset B$, then we say that A is a proper subset of B. It is easy to observe that $A = B$ if and only if $A \subseteq B$ and $B \subseteq A$.

Axiom of power set. *Let A be a set. Then there exists a set whose elements are precisely the subsets of A.*

If A is a set, then we denote the set of all subsets of A by $\mathcal{P}(A)$.

Axiom of pair. *Let A and B be sets. Then there exists a set C whose elements are precisely A and B.*

Let A be a set. Then by the axiom of pair, $\{A, A\}$ is a set, and by axiom of extension it is precisely $\{A\}$.

Let A and B be sets. Let $a \in A$ and $b \in B$. Then by the axiom of pair, $\{\{a\}, \{a, b\}\}$ is a set. We denote it by (a, b). The set (a, b) is called the ordered pair. We can easily note that $(a, b) = (c, d)$ if and only if $a = c$ and $b = d$. This indeed justifies the word "ordered pair".

Remark 1.3.1. The definition $(a, b) = \{\{a\}, \{a, b\}\}$ is called the Kuratowski definition of an ordered pair. There is another way of defining an ordered pair, which is called Hausdorff's definition. By Hausdorff's definition

$$(a, b) = \{\{a, \emptyset\}, \{b, \{\emptyset\}\}\}.$$

Axiom of union. *Let C be a set. Then there exists a set whose elements are precisely the elements of elements of C.*

Let C be a collection of sets. By $\bigcup C$ or $\bigcup\{A \mid A \in C\}$ we mean the set whose elements are precisely the elements of elements of C. This called the union of members of C. If $C = \{A, B\}$, then we denote the union by $A \cup B$.

Let C be a collection of sets. Then by the axiom of specification,

$$\bigcap C = \left\{ x \in \bigcup \,\middle|\, x \in A, \forall A \in C \right\}$$

is a set. This is called the intersection of members of C. If $C = \{A, B\}$, then we denote the intersection by $A \cap B$.

Let $F(x, y)$ denote a property such that for every x, there is a unique y for which $F(x, y)$ holds.

Axiom of replacement. *Let A be a set. Then there exists a set B such that for each $x \in A$, there is $y \in B$ such that $F(x, y)$ holds.*

Note that this axiom allows us to replace some of the elements of the set A by a set B.

Axiom of regularity. *Let A be a set. Then there exists a set B in the set A such that no member of B is in A.*

By this axiom we can prove that if A is a set, then $A \notin A$. On the contrary, suppose that there exists a set A such that $A \in A$. Then $A \in A \cap \{A\}$. This implies that $A \cap \{A\} \neq \emptyset$. This is a contradiction as by the axiom of regularity there is a set in $\{A\}$ whose intersection with $\{A\}$ is empty.

By this axiom we can also prove that if A and B are sets, then either $A \notin B$ or $B \notin A$. We leave this as an exercise.

Let A be a set. Then the successor of A is the set $A \cup \{A\}$. We denote it by A^+ or $A + 1$. A set S is called an inductive set if $\emptyset \in S$ and $A \in S$ implies that $A^+ \in S$.

Axiom of infinity. *There exists an inductive set.*

We can easily observe that an arbitrary intersection of inductive sets is again an inductive set. Thus the smallest inductive set exists. We denote it by $\mathbb{N}_0$.

Let the symbol 0 (zero) denote the empty set $\emptyset$. We denote 1 for the successor of 0, that is,

$$1 = 0^+ = 0 \cup \{0\} = \{0\} = \{\emptyset\}.$$

Similarly, 2 denotes the successor of 1, that is,

$$2 = 1^+ = 1 \cup \{1\} = \{0, 1\} = \{\emptyset, \{\emptyset\}\}.$$

Once we define $n-1$, we define n as the successor of $n-1$. Note that

$$n = \{0, 1, \ldots, n-1\}.$$

Thus

$$\mathbb{N}_0 = \{0, 1, \ldots, n, \ldots\}.$$

We denote by $\mathbb{N}$ the set $\mathbb{N}_0 \setminus \{0\}$, that is,

$$\mathbb{N} = \{1, 2, \ldots, n, \ldots\}.$$

The set $\mathbb{N}$ is called the set of natural numbers.

The above-mentioned nine axioms constitute the Zermelo–Fraenkel axiomatic system. Later, we will add one more axiom called the "Axiom of choice". We will assume all these ten axioms have a precise clarity, so that we are not going to face any problematic situation as far as the logic is concerned.

1.4 Relation

Let X and Y be sets. Then the Cartesian product $X \times Y$ of X and Y is the set of all ordered pairs (x, y), where $x \in X$ and $y \in Y$, that is,

$$X \times Y = \{(x, y) \mid x \in X \text{ and } y \in Y\}.$$

Note that $X \times Y \subseteq \mathcal{P}(\mathcal{P}(X \cup Y))$.

Definition 1.4.1. Let X and Y be sets. Then a relation R from X to Y is a subset of $X \times Y$.

Let $x \in X$ and $y \in Y$. If $(x, y) \in R$, then we say that x is related to y under the relation R. We also express it as $_xR_y$. A relation R from a set X to X is called a relation on X.

Example 1.4.2. Let X be a set. Then the relation $\Delta_X = \{(x, x) \mid x \in X\}$ is called the diagonal relation on X.

Example 1.4.3. Let X be a set. Then the relation

$$\in_X = \{(x, y) \mid x, y \in X \text{ and } x \in y\}$$

is called the membership relation on X.

Let R be a relation from a set X to a set Y. Then the set

$$\text{Dom}(R) = \{x \in X \mid \exists y \in Y \text{ such that } (x, y) \in R\}$$

is called the domain of R. Note that $\mathrm{Dom}(R) \subseteq X$. Also, the set

$$\mathrm{Im}(R) = \{y \in Y \mid \exists x \in X \text{ such that } (x,y) \in R\} \subseteq Y$$

is called the image or range of R. The set Y is called the codomain of R.

Definition 1.4.4. Let R be a relation from a set X to a set Y, and let S be a relation from Y to a set Z. Then the composition $S \circ R$ of relations R and S is defined as

$$S \circ R = \{(x,z) \in X \times Z \mid \exists y \in Y \text{ such that } (x,y) \in R \text{ and } (y,z) \in S\}.$$

We can show that if R, S, and T are the relations from a set X to a set Y, from Y to a set Z, and from Z to a set W, respectively, then $(T \circ S) \circ R = T \circ (S \circ R)$.

Definition 1.4.5. Let R be a relation from a set X to a set Y. Then the inverse relation R^{-1} of R is defined as

$$R^{-1} = \{(y,x) \in Y \times X \mid (x,y) \in X \times Y\}.$$

Example 1.4.6. Let $X = \{x,y,z\}$ and $R = \{(x,y),(x,z)\}$. Then $R^{-1} = \{(y,x),(z,x)\}$. Note that $R \circ R^{-1} = \{(y,y),(y,z),(z,y),(z,z)\}$ and $R^{-1} \circ R = \{(x,x)\}$.

Definition 1.4.7. A relation R on a set X is called
(i)   reflexive if $(x,x) \in R$ for all $x \in X$;
(ii)   symmetric if $(x,y) \in R$ implies $(y,x) \in R$ for all $x,y \in X$;
(iii)   transitive if $(x,y) \in R$ and $(y,z) \in R$ imply $(x,z) \in R$ for all $x,y,z \in X$.

Note that R is reflexive if and only if $\Delta_X \subseteq R$, R is symmetric if and only if $R^{-1} = R$, and R is transitive if and only if $R \circ R \subseteq R$.

Definition 1.4.8. A relation R on a set X is said to be an equivalence relation if it is reflexive, symmetric, and transitive.

Definition 1.4.9. Let R be an equivalence relation on a set X. Then the equivalence class $[x]$ of $x \in X$ is the set of all elements of X that are related to x. In other words,

$$[x] = \{y \in X \mid (x,y) \in R\}.$$

We also denote the equivalence class $[x]$ by R_x or $\bar{x}$. Since $x \in [x]$, $[x] \neq \emptyset$. We can easily show that the equivalence classes $[x]$ and $[y]$ are either equal or disjoint according to $(x,y) \in R$ or $(x,y) \notin R$, respectively. The set X/R of all equivalence classes of elements of a set X under an equivalence relation R is called the quotient set of X modulo R. In other words,

$$X/R = \{[x] \mid x \in X\}.$$

Definition 1.4.10. Let X be a set. A subset $\mathcal{A}$ of the power set of X is called a partition of X if the union of all members of $\mathcal{A}$ is X and any two distinct sets A and B in $\mathcal{A}$ are disjoint.

We can clearly note that if R is an equivalence relation on a set X, then the quotient set X/R is a partition of X. Conversely, given a partition $\mathcal{A}$ of a set X, we can define the equivalence relation

$$R = \{(x,y) \mid \exists A \in \mathcal{A} \text{ such that } x, y \in A\}.$$

We can also check that the quotient set X/R is precisely the partition $\mathcal{A}$.

Definition 1.4.11. A relation R on a set X is called
(i)   antisymmetric if whenever $(x,y) \in R$ and $(y,x) \in R$, then $x = y$.
(ii) asymmetric if $(x,y) \in R$ implies $(y,x) \notin R$ for all $x, y \in X$.
(iii) partial order if it is reflexive, antisymmetric, and transitive.
(iv) strict order if it is asymmetric and transitive.

If R is a partial order relation on a set X and $(x,y) \in R$, then we denote it by $x \leq y$ or $y \geq x$. The pair $(X, \leq)$ is called a partially ordered set or poset. When the partial order on X is understood, we will say that X is a partially ordered set. If S is a strict order on a set X and $(x,y) \in S$, then we denote it as $x < y$ or $y > x$. We can easily show that if S is a strict order on X, then the relation R on X defined by

$$(x,y) \in R \quad \text{if and only if} \quad (x,y) \in S \text{ or } x = y$$

is a partial order on X. Conversely, if R is a partial order on X, then the relation S on X defined by

$$(x,y) \in S \quad \text{if and only if} \quad (x,y) \in R \text{ or } x \neq y$$

is a strict order on X.

Example 1.4.12. Let $(X, \leq_1)$ and $(Y, \leq_2)$ be two posets. The relation $\leq$ on $X \times Y$ defined by $(x_1, y_1) \leq (x_2, y_2)$ if either $(x_1 \leq_1 x_2)$ or $(x_1 = x_2$ and $y_1 \leq_2 y_2)$ is a partial order on $X \times Y$. This is called the lexicographic order or dictionary order on $X \times Y$.

Example 1.4.13. Let $(X, \leq)$ be a poset and $Y \subseteq X$. Then $\leq \cap (Y \times Y)$ is a partial order on Y induced by the partial order on X.

We say that two elements x and y of a poset X are comparable if either $x \leq y$ or $y \leq x$. A partial ordered relation $\leq$ on a set X is called a total order if any two elements of X are comparable. A subset Y of a poset $(X, \leq)$ is called a chain if Y is a total ordered set with respect to the induced partial order on Y.

Let A be a subset in a poset X. Then $x \in X$ is called an upper bound (respectively, a lower bound) of A if $a \leq x$ (respectively, $x \leq a$) for all $a \in A$. It is not necessary that an upper bound or a lower bound of a set A exists in a poset X. If an upper bound (respectively, a lower bound) of a set A exists in X, say x, then we say that A is bounded above (respectively, bounded below) by x. An element $m \in X$ is called a maximal (respectively, a minimal) of X if for all elements $x \in X$ with $m \leq x$ (respectively, $x \leq m$) we have that $x = m$. An element $g \in X$ (respectively, $s \in X$) is called the greatest or largest element (respectively, the smallest or least element) of X if $x \leq g$ (respectively, $s \leq x$) for all $x \in X$. Note that the greatest or the least element (if it exists) is unique. A partial order $\leq$ on a set X is called a well order if every nonempty subset of X has the least element in it. If $\leq$ is a well order on X, then we say that $(X, \leq)$ is a well-ordered set.

Let A be a subset of a poset X. Then $l \in X$ is called the least upper bound or supremum of A if it is an upper bound of A and the least element in the set of all upper bounds of A. We denote the least upper bound of a set A (if it exists) by lub(A) or sup(A). Also, an element $g \in X$ is called the greatest lower bound or infimum of A if it is a lower bound of A and the greatest element in the set of all lower bounds of A. We denote the greatest lower bound of a set A (if it exists) by glb(A) or inf(A). A partial order $\leq$ on a set X is called a complete order if all the nonempty subsets of X that are bounded above have the least upper bounds in X. If a partial order $\leq$ is a complete order on X, then we also say that X has the least upper bound property. We can show that a partial order $\leq$ on a set X is a complete order if and only if all the nonempty subsets of X that are bounded below have the greatest lower bounds in X.

1.5 Map

The notion of a map is important in mathematics in a lot of ways. One of the most important when we try to figure out if two objects are the same in some sense. Let us understand it by the following nonmathematical example.

Example 1.5.1. Ms. Vinny has two kids called Mr. Vanjual and Mr. Venu. She often takes help from these kids while working in her kitchen. Mr. Vanjul knows only 1, 2, and 3, whereas Mr. Venu knows only a, b, and c. She has three boxes containing sugar, salt, and baking soda, respectively, in the three boxes. She puts the labels 1, 2, and 3, respectively, for sugar, salt, and baking soda. She also puts the labels a, b, and c, respectively, for sugar, salt, and baking soda. Whenever she needs sugar with help of Mr. Vanjul, she just says "Please give me the box having label 1" and similarly for the other labels. In this way, we can say that the set $\{1, 2, 3\}$ is the same as the set $\{a, b, c\}$ as far as this particular purpose is concerned.

Definition 1.5.2. A relation f from a set X to a set Y is called a map or function if for every $x \in X$, there is a unique $y \in Y$ such that $(x, y) \in f$. In other words,

(i) for all $x \in X$, there is $y \in Y$ such that $(x, y) \in f$, and

(ii) if $(x, y_1) \in f$ and $(x, y_2) \in f$, then $y_1 = y_2$.

If f is a map from a set X to a set Y and $(x, y) \in f$, then we say that y is the image of x under f. We denote it as $y = f(x)$. By the notation $f : X \to Y$ we mean that f is a map from X to Y. Two maps $f, g : X \to Y$ are called equal if $f(x) = g(x)$ for all $x \in X$.

Example 1.5.3. There is only one map $\emptyset$ from the empty set to any set. This map is called the empty map. There is no map from a nonempty set to the empty set.

Example 1.5.4. The diagonal relation Δ_X is a map on a set X. This map is called the identity map. We denote it by I_X.

Example 1.5.5. Let A be a subset of a set X. Then the relation $i_A = \{(a, a) \mid a \in A\}$ is a map from A to X. This map is called the inclusion map. We also denote it by $A \hookrightarrow X$.

Example 1.5.6. Let X and Y be sets and fix $c \in Y$. Then the relation $f = X \times \{c\}$ is a map from X to Y. This map is called a constant map. Thus $f(x) = c$ for all $x \in X$.

Example 1.5.7. Let X_1 and X_2 be sets. Then $p_i : X_1 \times X_2 \to X_i$ $(i = 1, 2)$ defined by $p_i(x_1, x_2) = x_i$ is a map. This map is called the ith projection.

Example 1.5.8. Let R be an equivalence relation on a set X. Then the correspondence $\nu_R : X \to X/R$ defined by $\nu_R(x) = [x]$ is a map. This map is called the quotient map.

Example 1.5.9. Let $f : X \to Y$ and $g : Y \to Z$ be maps. Then their composition $g \circ f$ as a relation is a map from X to Z. Note that $(g \circ f)(x) = g(f(x))$.

Example 1.5.10. Let $f : X \to Y$ be a map, and let $A \subseteq X$. Then the map $f \circ i_A : A \to Y$ is called the restriction of f on A. We denote this map by $f \mid_A$.

Definition 1.5.11. Let $f : X \to Y, g : Y \to Z$, and $h : X \to Z$ be maps. Then the diagram

$$X \xrightarrow{f} Y$$
$$h \searrow \quad \downarrow g$$
$$Z$$

is called commutative if $h = g \circ f$.

Definition 1.5.12. A map $f : X \to Y$ is called:

(i) Injective or one-one if $f(x) = f(y)$ implies $x = y$ for all $x, y \in X$.

(ii) Surjective or onto if for all $y \in Y$, there is $x \in X$ such that $y = f(x)$.

(iii) Bijective if it is injective and surjective.

Let $f : X \to Y$ be a map. Then the inverse relation f^{-1} from Y to X need not be a map. We can show that f^{-1} is a map if and only if f is bijective. In this case, f^{-1} is also bijective.

Let $f : X \to Y$ be a map, and let A be a subset of X. Then the set

$$f(A) = \{f(a) \mid a \in A\} \subseteq Y$$

is called the image of A under f. Note that f is surjective if and only if $f(X) = Y$. We can observe that if A and B are subsets of X, then $f(A \cup B) = f(A) \cup f(B)$ and $f(A \cap B) \subseteq f(A) \cap f(B)$. Also, $f(A \cap B) = f(A) \cap f(B)$ for all subsets A and B of X if and only if f is injective.

Let $f : X \to Y$ be a map, and let B be a subset of Y. Then the set

$$f^{-1}(B) = \{x \in X \mid f(x) \in B\} \subseteq X$$

is called the inverse image or preimage of B under f. We can show that if A and B are subsets of Y, then $f^{-1}(A \cup B) = f^{-1}(A) \cup f^{-1}(B)$, $f^{-1}(A \cap B) = f^{-1}(A) \cap f^{-1}(B)$, and $f^{-1}(A \setminus B) = f^{-1}(A) \setminus f^{-1}(B)$.

Let I and X be sets. Then the domain I of a surjective map $f : I \to X$ is called an indexing set. The use of an indexing set is to collect as many sets (which are not necessarily distinct) as elements in I. Let

$$S = \{A_\alpha \mid \alpha \in I, \text{where } I \text{ is an indexing set}\}$$

be a collection of sets. The union of the members of the above collection is denoted by $\bigcup_{\alpha \in I} A_\alpha$ or $\bigcup\{A_\alpha \mid \alpha \in I\}$. We have the corresponding notation $\bigcap_{\alpha \in I} A_\alpha$ for the intersection. We can generalize the behavior of arbitrary union or intersection of subsets under the map $f : X \to Y$; for example,

$$f\left(\bigcup_{\alpha \in I} A_\alpha\right) = \bigcup_{\alpha \in I} f(A_\alpha).$$

We leave the others as an exercise for the reader. We are now able to have one more axiom, "Axiom of choice". This axiom is independent of all the axioms of the Zermelo–Fraenkel system.

Axiom of choice. *Let X be a set of nonempty sets. Then there is a function $f : X \to \bigcup X$ such that $f(A) \in A$ for all $A \in X$. Such a function f is called a choice function.*

Let I be an indexing set, and let $\mathcal{C} = \{X_\alpha \mid \alpha \in I\}$ be a collection of sets. Then the set

$$\prod_{\alpha \in I} X_\alpha = \left\{x : I \to \bigcup_{\alpha \in I} X_\alpha \ \middle|\ x(\alpha) \in X_\alpha\right\}$$

is called the Cartesian product of the collection $\mathcal{C}$.

Let $f : X \to Y$ be a map. Suppose that R and S are equivalence relations on X and Y, respectively. It is natural to think of a map between the quotient sets X/R and Y/S such that the equivalence class of $x \in X$ corresponds to the equivalence class of $f(x) \in Y$. In general, this correspondence need not be a map.

For a map $f : X \to Y$, we define the map $(f \times f) : X \times X \to Y \times Y$ by $(f \times f)(x_1, x_2) = (f(x_1), f(x_2))$. Suppose that R and S are equivalence relations on X and Y, respectively. Then $f : X \to Y$ is called a relation-preserving if $(f \times f)(R) \subseteq S$. In other words, $(f \times f)(x_1, x_2) \in S$ for all $(x_1, x_2) \in R$.

If $f : X \to Y$ be a map, then the equivalence relation $\text{Ker} f = (f \times f)^{-1}(\Delta_Y)$ on the set X is called the kernel of f. Note that f is injective if and only if $\text{Ker} f = \Delta_X$.

Let R and S be equivalence relations on the sets X and Y, respectively. Let $f : X \to Y$ be a relation-preserving map. Then we have the map $\tilde{f} : X/R \to Y/S$ defined by $\tilde{f}([x]) = [f(x)]$. Observe that the following diagram is commutative:

$$
\begin{array}{ccc}
X & \xrightarrow{\ f\ } & Y \\
{\scriptstyle \nu_R}\downarrow & & \downarrow{\scriptstyle \nu_S} \\
X/R & \xrightarrow{\ \tilde{f}\ } & Y/S
\end{array}
$$

We can check that a map $\tilde{f}$ that makes the above diagram commutative is unique. Also, if two maps $f : X \to Y$ and $\tilde{f} : X/R \to Y/S$ make the above diagram commutative, then f is necessarily a relation-preserving map.

Let us take $S = \Delta_Y$ in the above diagram and suppose $R \subseteq \text{Ker} f$. Then we have the map $g : X/R \to Y$ defined by $g([x]) = f(x)$. Note that $g \circ \nu_R = f$ and such a map is unique. Observe that to have a well-defined map $g : X/R \to Y$, it is necessary that $R \subseteq \text{Ker} f$. In this case the relation $(f \times f)(R)$ is an equivalence relation on Y. Moreover, if $f : X \to Y$ is surjective and $R \subseteq \text{Ker} f$, then we have a unique bijective map $\tilde{f} : X/R \to Y/S$ that makes the following diagram commutative:

$$
\begin{array}{ccc}
X & \xrightarrow{\ f\ } & Y \\
{\scriptstyle \nu_R}\downarrow & & \downarrow{\scriptstyle \nu_S} \\
X/R & \xrightarrow{\ \tilde{f}\ } & Y/S
\end{array}
$$

We are now in position to prove the following result, which is fundamental in nature.

Theorem 1.5.13 (Fundamental theorem of maps). *Let R be an equivalence relation on a set X, and let $f : X \to Y$ be a map. Then there is a unique map $\tilde{f} : X/R \to Y$ that makes the following diagram commutative if and only if $R \subseteq \text{Ker} f$:*

$$
\begin{array}{ccc}
X & \xrightarrow{\ f\ } & Y \\
{\scriptstyle \nu_R}\downarrow & \nearrow{\scriptstyle \tilde{f}} & \\
X/R & &
\end{array}
$$

Also, $\tilde{f}$ is injective if and only if $R = \text{Ker} f$, and $\tilde{f}$ is surjective if and only if f is surjective.

Proof. The existence of such a map $\tilde{f} : X/R \to Y$ is guaranteed by the above discussion. Now

$$\operatorname{Ker}\tilde{f} = \{([x],[y]) \in X/R \times X/R \mid \tilde{f}([x]) = \tilde{f}([y])\}$$
$$= \{([x],[y]) \in X/R \times X/R \mid (x,y) \in \operatorname{Ker}f\}.$$

Note that $\tilde{f}$ is injective if and only if $\operatorname{Ker}\tilde{f} = \Delta_{X/R}$ if and only if $R = \operatorname{Ker}f$.

Since v_R is surejective and $\tilde{f} \circ v_R = f, \tilde{f}$ is surjective if and only if f is surjective. □

1.6 Construction of number systems. I

In this section, we provide the construction of some number systems. Recall that

$$\mathbb{N}_0 = \{0, 1, \ldots, n, \ldots\}.$$

Also, recall from the construction of $\mathbb{N}_0$ that $n \in n + 1$ for all $n \in \mathbb{N}_0$. Note that the following properties hold in the set $\mathbb{N}_0$:

(i) $0 \in \mathbb{N}_0$.
(ii)   if $n \in \mathbb{N}_0$, then $n^+ \in \mathbb{N}_0$.
(iii) $n^+ = m^+ \Leftrightarrow n = m$ for all $n, m \in \mathbb{N}_0$.
(iv) $0 \neq n^+$ for all $n \in \mathbb{N}_0$.
(v)   Suppose that $P(m)$ denotes some property that involves $m \in \mathbb{N}_0$. If $P(0)$ holds and $P(n^+)$ holds whenever $P(n)$ holds, then $P(n)$ holds for all $n \in \mathbb{N}_0$.

We will only prove (v). Let $A = \{n \in \mathbb{N}_0 \mid P(n) \text{ holds}\}$. Note that A is an inductive set. Since $\mathbb{N}_0$ is the smallest inductive set, $\mathbb{N}_0 \subseteq A$. Since $A \subseteq \mathbb{N}_0$, $A = \mathbb{N}_0$. Property (v) is called the principle of mathematical induction. Using this, we can prove that if $P(1)$ holds and $P(n^+)$ holds whenever $P(n)$ holds, then $P(n)$ holds for all $n \in \mathbb{N}$. We will also call this the principle of mathematical induction.

Definition 1.6.1. Let X be a nonempty set. Then a binary operation on X is a map $\circ : X \times X \to X$. We denote the element $\circ(x,y)$ of X by $x \circ y$.

Having defined the natural numbers, we wish to define two binary operations on $\mathbb{N}_0$, the addition and multiplication. To define these binary operations, we need the following:

Theorem 1.6.2 (Recursion theorem). *Let $a \in \mathbb{N}_0$ be a fixed element, and let $f : \mathbb{N}_0 \to \mathbb{N}_0$ be a map. Then there is a unique map $g : \mathbb{N}_0 \to \mathbb{N}_0$ such that $g(0) = a$ and $g(n^+) = f(g(n))$ for all $n \in \mathbb{N}_0$.*

Proof. Consider the set S consisting of the relations h on $\mathbb{N}_0$ such that $(0, a) \in h$ and if $(n, m) \in h$, then $(n^+, f(m)) \in h$. Note that $S \neq \emptyset$ as $\mathbb{N}_0 \times \mathbb{N}_0 \in h$. We can check that $g = \bigcap S$ is the required map and such a map is unique. □

Define the map $\psi : \mathbb{N}_0 \to \mathbb{N}_0$ by $\psi(n) = n^+$. Let $m \in \mathbb{N}_0$ be fixed. By the recursion theorem we have a unique map $g_m : \mathbb{N}_0 \to \mathbb{N}_0$ defined by $g_m(0) = m$ and $g_m(n^+) = \psi(g_m(n))$.

Let $m, n \in \mathbb{N}_0$. We define the binary operation + (called the addition) on the set $\mathbb{N}_0$ by $m + n = g_m(n)$. Observe that $m + 0 = g_m(0) = m$ and $m + n^+ = g_m(n^+) = \psi(g_m(n)) = (m+n)+1$. Therefore the addition is defined on $\mathbb{N}_0$ recursively. Also, note that $n+1 = n^+$.

From the above discussion we observe that for $m \in \mathbb{N}_0$, we have the map $\psi_m : \mathbb{N}_0 \to \mathbb{N}_0$ defined by $\psi_m(n) = n + m$. By the recursion theorem we have a unique map $\mu_m : \mathbb{N}_0 \to \mathbb{N}_0$ defined by $\mu_m(0) = 0$ and $\mu_m(n^+) = \psi_m(\mu_m(n))$.

Let $m, n \in \mathbb{N}_0$. We define the binary operation · (called the multiplication) on the set $\mathbb{N}_0$ by $m \cdot n = \mu_m(n)$. In place of $m \cdot n$, we will only write mn. Observe that $m \cdot 0 = \mu_m(0) = 0$ and $mn^+ = \mu_m(n^+) = \psi_m(\mu_m(n)) = (mn) + m$. Therefore the multiplication is defined on $\mathbb{N}_0$ recursively. Observe that the multiplication on $\mathbb{N}_0$ is the repeated addition.

Let $m, n \in \mathbb{N}_0$. We say that $n < m$ if $n \in m$. Observe that $<$ is a strict order on $\mathbb{N}_0$, and hence we get a partial order $\leq$ on $\mathbb{N}_0$. Also observe that $(\mathbb{N}_0, \leq)$ is a well-ordered set. This is usually called the well-ordering principle. Clearly, given $m, n \in \mathbb{N}_0$, we have either $m < n$ or $n < m$ or $m = n$, and no two of these can occur together. This is called the trichotomy law. It is easy to observe that $<$ on $\mathbb{N}_0$ respects the addition and multiplication in the sense that if $m, n > 0$, then $m + n > 0$ and $mn > 0$. By the statement "$\mathbb{N}_0$ is a number system" we mean the set $\mathbb{N}_0$ with two binary operations + and · and a partial order $\leq$ that respects these binary operations. Finally, observe that $m < n$ if and only if there is unique $r \in \mathbb{N}$ such that $n = m + r$.

Let us try to solve the equation $X + 2 = 1$ in the number system $\mathbb{N}_0$. Suppose there is $a \in \mathbb{N}_0$ such $a + 2 = 1$. This implies that $a + 1 = 0$. This is a contradiction since 0 is not successor of any element in $\mathbb{N}_0$. It is natural to search where the solution of the equation $X + a = b$ exists, where $a, b \in \mathbb{N}_0$. For this, we define the relation R on the set $\mathbb{N}_0 \times \mathbb{N}_0$ by $((a, b), (c, d)) \in R$ if $a + d = b + c$. We can check that R is an equivalence relation on $\mathbb{N}_0 \times \mathbb{N}_0$. The quotient set $(\mathbb{N}_0 \times \mathbb{N}_0)/R$ is called the set of integers, and we denote it by $\mathbb{Z}$. The element $x = [(a, b)] \in \mathbb{Z}$ is called an integer.

Let $x = [(a, b)]$ and $y = [(c, d)]$ be integers. We define the addition and multiplication in $\mathbb{Z}$ as follows:

$$[(a, b)] + [(c, d)] = [(a + c, b + d)],$$
$$[(a, b)][(c, d)] = [(ac + bd, ad + bc)].$$

We can check that the addition and multiplication are binary operations on $\mathbb{Z}$ and that $(\mathbb{Z}, +, \cdot)$ is a ring. Note that $0 = [(a, a)]$ is the additive identity of $(\mathbb{Z}, +)$ and that for $x = [(a, b)] \in \mathbb{Z}$, $-x = [(b, a)]$ is the additive inverse of x.

Let $x = [(a, b)]$ and $y = [(c, d)]$ be integers. We say that $x < y$ if $a + d < b + c$. We can check that $<$ is a strict order on $\mathbb{Z}$, which respects the addition and multiplication. Define the map $f : \mathbb{N}_0 \to \mathbb{Z}$ by $f(n) = [(n, 0)]$. Check that f is injective and preserves the addition, multiplication, and strict order $<$ on $\mathbb{Z}$ in the sense that $f(n+m) = f(n)+f(m)$, $f(nm) = f(n)f(m)$, and $n < m \Rightarrow f(n) < f(m)$. We call such a map an embedding.

Let $x = [(a,b)]$. Then $a > b$ or $a = b$ or $a < b$. If $a = b$, then $x = 0$. If $a > b$, then $[(a,b)] = [(a-b,0)]$. If $a < b$, then $[(a,b)] = [(0,b-a)]$. Let $n \in \mathbb{N}$. In the sense of the above embedding, we represent $[(n,0)]$ by n and $[(0,n)]$ by $-n$. In this way, we write $\mathbb{Z} = \mathbb{N} \cup \{0\} \cup (-\mathbb{N})$.

Let $a,b \in \mathbb{Z}$ with $b \neq 0$. Observe that there are equations $bX = a$ that are not solvable in the set of integers, for example, the equation $2X = 1$. In the same spirit as before, we will enlarge the system of integers. Define the relation R on the set $\mathbb{Z} \times (\mathbb{Z} \setminus \{0\})$ by $((a,b),(c,d)) \in R$ if $ad = bc$. Clearly, R is an equivalence relation on $\mathbb{Z} \times (\mathbb{Z} \setminus \{0\})$. The equivalence class $[(a,b)]$ is called a rational number. We denote it by $\frac{a}{b}$. The quotient set $\mathbb{Z} \times (\mathbb{Z} \setminus \{0\})/R$ is denoted by $\mathbb{Q}$.

Let $\frac{a}{b}, \frac{c}{d} \in \mathbb{Q}$. We define the addition and multiplication in $\mathbb{Q}$ as follows:

$$\frac{a}{b} + \frac{c}{d} = \frac{ad+bc}{bd},$$
$$\frac{a}{b}\frac{c}{d} = \frac{ac}{bd}.$$

We can check that $(\mathbb{Q}, +, \cdot)$ is a field.

Definition 1.6.3. A field $(\mathbb{F}, +, \cdot)$ is called an ordered field if there is a strict order $<$ on $\mathbb{F}$ such that
(i)   given $x,y \in \mathbb{F}$, one and only one of the following is satisfied:

$$x < y \quad \text{or} \quad x = y \quad \text{or} \quad y < x;$$

(ii)   if $x < y$, then for all $z \in \mathbb{F}$, we have $x + z < y + z$;
(iii)   if $x > 0$ and $y > 0$, then $xy > 0$.

We define $<$ on $\mathbb{Q}$ as $\frac{a}{b} < \frac{c}{d}$ if $ad < bc$. Clearly, $(\mathbb{Q}, +, \cdot)$ is an ordered field with this strict order $<$ on $\mathbb{Q}$. The map $f : \mathbb{Z} \to \mathbb{Q}$ by $f(k) = \frac{k}{1}$ is an embedding of $\mathbb{Z}$ in $\mathbb{Q}$.

1.7 Construction of number systems. II

In the previous section, we enlarged the number system by searching the solutions of some equations. We can easily show that the equation $X^2 = 2$ has no solution in $\mathbb{Q}$. In the same temperament, we may be interested in enlarging the rational number system. Indeed, we will do something more. We will see that the rational number system is incomplete in some sense, and we will complete it by getting an enlarged system, which is unique in some sense.

Consider the set

$$L = \{x \in \mathbb{Q} \mid x > 0 \text{ and } x^2 < 2\}.$$

Note that L is a nonempty subset of $\mathbb{Q}$ since, for example, $1 \in L$. Note that if $x \in L$, then $x^2 < 2 < 4$. This implies that $x < 2$. Therefore L is bounded above by 2. We claim that the least upper bound of L does not exist in $\mathbb{Q}$. On the contrary, suppose that $l = \text{lub}(L) \in \mathbb{Q}$. Clearly, $l > 0$ as $1 \in L \Rightarrow 1 \leq l$. Then we have one of the following:

$$l^2 < 2 \quad \text{or} \quad l^2 = 2 \quad \text{or} \quad l^2 > 2.$$

Since there is no rational whose square is 2, $l^2 < 2$ or $l^2 > 2$. First, suppose that $l^2 < 2$. Then $l \in L$. Now we will show that we can find a larger rational number $l + h$ $(h > 0)$ such that $(l + h)^2 < 2$. Suppose that $0 < h < 1$. Then

$$(l + h)^2 = l^2 + 2lh + h^2 < l^2 + 2lh + h.$$

Therefore it is sufficient to find h such that $l^2 + 2lh + h = 2$. This gives $h = \frac{2 - l^2}{2l + 1}$. For this h, $l + h \in L$. This shows that l cannot be the least upper bound of L. This is a contradiction.

Now suppose that $l^2 > 2$. We will show that we can find a smaller rational number $l - k$ $(k > 0)$ such that $(l - k)^2 > 2$. Note that

$$(l - k)^2 = l^2 - 2lk + k^2 > l^2 - 2lk.$$

Therefore it is sufficient to find k such that $l^2 - 2lk = 2$. This gives $k = \frac{l^2 - 2}{2l}$. This shows that l cannot be the least upper bound of L. This is again a contradiction.

From the above discussion we conclude that $\mathbb{Q}$ is not a complete ordered set. Now we enlarge the ordered field $\mathbb{Q}$ to an ordered field that is complete with respect to the order defined on this enlarged system. We will call this as a complete ordered field. We can note that $\mathbb{Q}$ is partitioned into the following two sets:

$$L = \{x \in \mathbb{Q} \mid x > 0 \text{ and } x^2 < 2\} \cup \{x \in \mathbb{Q} \mid x \leq 0\},$$
$$U = \{x \in \mathbb{Q} \mid x > 0 \text{ and } x^2 > 2\}.$$

Definition 1.7.1. A pair (L, U) of nonempty subsets of $\mathbb{Q}$ is called a Dedekind cut if the following properties hold:
(i) $L \cup U = \mathbb{Q}$ and $L \cap U = \emptyset$,
(ii) If $x \in L$ and $y \in U$, then $x < y$.
(iii) The set L does not contain the largest rational number.

Remark 1.7.2. We could define the Dedekind cut by replacing condition (iii) by "The set U does not contain the smallest rational number".

Definition 1.7.3. The set of all Dedekind cuts is denoted by $\mathbb{R}$. If $\alpha = (L, U) \in \mathbb{R}$, then α is called a real number.

Remark 1.7.4. There are two types of Dedekind cuts. The first type of Dedekind cuts consists of those (L, U) in which U contains the smallest rational number. The second type consists of those (L, U) in which U does not contain the smallest rational number. Later, we will observe that the first type determines the rationals and the other type determines some extra elements, which we will call the irrationals. These are precisely to fill the gaps of the rationals.

Let $\alpha = (L_1, U_1)$ and $\beta = (L_2, U_2)$ be two real numbers. We say that $\alpha = \beta$ if $L_1 = L_2$. This will clearly imply $U_1 = U_2$. We also say that $\alpha < \beta$ if L_1 is properly contained in L_2. We can check that $<$ is a strict order on $\mathbb{R}$ and one and only one of the following holds:

$$\alpha < \beta \quad \text{or} \quad \alpha = \beta \quad \text{or} \quad \beta < \alpha.$$

Proposition 1.7.5. *Let $\alpha = (L, U)$ be a real number, and let $d \in \mathbb{Q}$ with $d > 0$. Then there exist $x \in L$ and $y \in U$ such that $y - x = d$.*

Proof. First, note that either $d \in L$ or $d \in U$. Suppose that $d \in L$. If $z \in U$, then $d < z$. It is easy to observe that there is $n \in \mathbb{N}$ such that $nd > z$. Therefore $nd \in U$. Consider the following finite set

$$A = \{kd \mid k \in \mathbb{N} \text{ and } 1 \le k \le n\}.$$

Observe that there exist two consecutive members of A, say $x = ld$ and $y = (l + 1)d$, such that $x \in L$ and $y \in U$. Note that $y - x = d$.

Now suppose that $d \in U$. If $0 \in L$, then we take $x = 0$ and $y = d$. If $0 \in U$, then by the similar argument as before we can find $l \in \mathbb{N}$ such that $x = -ld \in L$ and $y = -ld + 1 \in U$. ∎

Let $\alpha = (L_1, U_1)$ and $\beta = (L_2, U_2)$ be real numbers. We define the addition $\alpha + \beta$ of α and β as (L, U), where $L = L_1 + L_2 = \{a + b \mid a \in L_1, b \in L_2\}$ and $U = \mathbb{Q} \setminus L$. Now we show that (L, U) is a Dedekind cut.

Clearly, L and U partition $\mathbb{Q}$. Let $x \in L$ and $y \in U$. On the contrary, suppose that $y \le x$. Since $x \in L$, $x = a + b$, where $a \in L_1$, $b \in L_2$. This implies that $y - b \le a$. Therefore $y - b \in L_1$. This shows that $y = y - b + b \in L$. This is a contradiction. Since L_1 and L_2 do not contain the largest rational, L does not contain the largest rational.

We can easily check that if α, β, and γ are real numbers, then $\alpha + (\beta + \gamma) = (\alpha + \beta) + \gamma$ and $\alpha + \beta = \beta + \alpha$.

Let $\mathbf{0} = (N, P)$, where $N = \{x \in \mathbb{Q} \mid x < 0\}$ and $P = \{x \in \mathbb{Q} \mid x \ge 0\}$. We can check that (N, P) is a Dedekind cut. From now on in this section, we will denote by N the set of negative rational numbers. We claim that $\alpha + \mathbf{0} = \alpha$ for every real number $\alpha = (L, U)$.

Let $a + \mathbf{0} = (L_1, U_1)$. Let $x \in L_1$. Then $x = a + b$, where $a \in L$ and $b \in N$. This implies that $x = a + b < a$. Therefore $x \in L$. Conversely, let $y \in L$. Since L has no largest element, there is $z > y$ such that $z \in L$. Then there is positive rational number k such that $z = y + k$. Equivalently, $y = z + (-k) \in L + N = L_1$. Hence $L = L_1$. This shows that $a + \mathbf{0} = a$.

Let $a = (L, U)$ be a real number. We define $-a = (L_1, U_1)$, where

$$L_1 = \{-x \mid x \in U, \text{ and } x \text{ is not the smallest element of } U \text{ (if exists)}\},$$
$$U_1 = \mathbb{Q} \setminus L_1.$$

Note that (L_1, U_1) is a Dedekind cut. We claim that $a + (-a) = \mathbf{0}$. Let $a + (-a) = (L_2, U_2)$. Let $x \in L_2$. Then $x = a + b$, where $a \in L$ and $b \in L_1$. By the definition, $b = -c$, where $c \in U$. By the property of Dedekind cut, $a < c = -b$, that is, $x = a + b < 0$. Therefore $x \in N$. Conversely, let $y \in N$. By Proposition 1.7.5 there are $u \in L$ and $v \in U$ such that $v - u = -y$. Then $y = u + (-v) \in L_2$. Hence $L_2 = N$. This shows that $a + (-a) = \mathbf{0}$.

Let $a = (L_1, U_1) > \mathbf{0}$ and $\beta = (L_2, U_2) > \mathbf{0}$. We define the product $a\beta$ of a and β as (L, U), where

$$L = N \cup \{xy \mid x \geq 0, y \geq 0, x \in L_1, y \in L_2\},$$
$$U = \mathbb{Q} \setminus L.$$

We first show that (L, U) is a Dedekind cut. Clearly, $L \neq \emptyset$. By Proposition 1.7.5 there are $x \in L_1$ and $y \in L_2$ such that $x + 1 \in U_1$ and $y + 1 \in U_2$. We claim that $(x + 1)(y + 1) \in U$. On the contrary, suppose that $(x + 1)(y + 1) \in L$. Then $(x + 1)(y + 1) = ab$, where $a \in L_1$ and $b \in L_2$. Note that $a > 0$ and $b > 0$. By the property of Dedekind cut, $a < x + 1$ and $b < y + 1$. This shows that $ab < (x + 1)(y + 1)$. This is a contradiction. Also, we can easily note that L and U partition $\mathbb{Q}$.

Since L_1 and L_2 have no largest elements, L has no largest element. Let $x \in L$ and $y \in U$ with $x > 0$ and $y > 0$. On the contrary, suppose that $y < x$ as $x \neq y$. Then $x = yk$ for some rational number $k > 1$. Since $x \in L$, $x = ab$, where $a \in L_1$ and $b \in L_2$. Note that $\frac{a}{k} \in L_1$; otherwise, $\frac{a}{k} \in U_1$, which would imply $a < \frac{a}{k}$. Now

$$y = \frac{x}{k} = \frac{a}{k}b \in L.$$

This is a contradiction.

Now we define the product of any two real numbers as follows:

$$a\beta = \begin{cases} -(-a)\beta & \text{if } a < \mathbf{0}, \\ -a(-\beta) & \text{if } \beta < \mathbf{0}, \\ (-a)(-\beta) & \text{if } a < \mathbf{0}, \beta < \mathbf{0}. \end{cases}$$

We can easily check that $a(\beta\gamma) = (a\beta)\gamma$ and $a\beta = \beta a$ for all real numbers a, β, and γ.

Let $\alpha = (L, U) > \mathbf{0}$. We define $\alpha^{-1} = (L_1, U_1)$, where

$$L_1 = N \cup \{0\} \cup \{x^{-1} \mid x \in U, \text{ and } x \text{ is not the smallest element of } U \text{ (if exists)}\},$$
$$U_1 = \mathbb{Q} \setminus L_1.$$

We claim that (L_1, U_1) is a Dedekind cut. Clearly, L_1 and U_1 partition $\mathbb{Q}$. Let $z \in L_1$ and $z > 0$. Then $z = x^{-1}$, where $x \in U$. Now there exists $y \in U$ such that $y < x$. Then $y^{-1} \in L_1$ and $z < y^{-1}$. This shows that L_1 does not have the largest element.

Let $x \in L_1$ and $y \in U_1$. We claim that $x < y$. On the contrary, suppose that $y < x$ as $x \neq y$. Then $x = yk$ for some rational $k > 1$. Since $x \in L_1$, $x = z^{-1}$, where $z \in U$. Clearly, $zk \in U$. Then $y = xk^{-1} = (zk)^{-1} \in L_1$. This is a contradiction.

Let $\mathbf{1} = (A, B)$, where $A = \{x \in \mathbb{Q} \mid x < 1\}$ and $B = \{x \in \mathbb{Q} \mid x \geq 1\}$. Note that $\mathbf{1}$ is a real number. We can easily show that $\mathbf{1}\alpha = \alpha$ for every real number $\alpha = (L, U)$.

Let $\alpha = (L_1, U_1) > \mathbf{0}$ and $\alpha^{-1} = (L_2, U_2)$. We claim that $\alpha\alpha^{-1} = \mathbf{1}$. For this, let $\alpha\alpha^{-1} = (L, U)$. We have to show that $L = A$.

Let $x \in L$. If $x \leq 0$, then $x \in A$. Therefore we can suppose that $x = ab$, where $a \in L_1$, $b \in L_2$, $a, b > 0$, and $b \in L_2$. Then $b = c^{-1}$, where $c \in U_1$. This implies that $a < c$. Then $x = ab = ac^{-1} < 1$. Hence $x \in A$.

Conversely, let $x \in A$. If $x \leq 0$, then $x \in L$. Suppose that $0 < x < 1$. Choose $u \in U_2$. Choose $n \in \mathbb{N}$ such that $n(1 - x) > u$. Now choose $a \in L_1$ and $b, c \in U_1$ with $c < b$ such that $b - a < \frac{1}{n}$. Note that $c > 0$. Then $b^{-1} < c^{-1}$. This implies that $b^{-1} \in L_2$. Since $b - a < \frac{1}{n}$, $ab^{-1} > 1 - \frac{b^{-1}}{n}$. Since $u \in U_2$ and $b^{-1} \in L_2$, $b^{-1} < u$. Then

$$x < 1 - \frac{u}{n} < 1 - \frac{b^{-1}}{n} < ab^{-1}.$$

Since $a \in L_1$ and $b^{-1} \in L_2$, $x \in L$. Thus $L = A$.

If $\alpha < \mathbf{0}$, then we define $\alpha^{-1} = -(-\alpha)^{-1}$. We can easily check that $\alpha\alpha^{-1} = \mathbf{1}$ in this case also.

Now we show that $\alpha(\beta + \gamma) = \alpha\beta + \alpha\gamma$ for all real numbers $\alpha = (L_1, U_1)$, $\beta = (L_2, U_2)$, and $\gamma = (L_3, U_3)$. It is sufficient to prove it for positive real numbers. Let $\alpha(\beta + \gamma) = (L, U)$ and $\alpha\beta + \alpha\gamma = (L', U')$. We have to show that $L = L'$. We will only prove that all positive elements of L are in L' and all positive elements of L' are in L. Let $x \in L$ and $x > 0$. Since $x = a(b + c) = ab + ac$, $x \in L'$. Conversely, suppose that $x \in L'$ and $x > 0$. Then $x = ab + cd$ with the understanding where $a, b, c,$ and d belong. If $a = c$, then $x \in L$. Suppose that $a \neq c$. We can suppose that $c < a$. Then $\frac{c}{a}d < d$. This shows that $\frac{c}{a}d \in L_3$. Therefore $x = ab + cd = a(b + \frac{c}{a}d) \in L$. Hence $L = L'$.

Thus we have shown the following:

Theorem 1.7.6. *The system* $(\mathbb{R}, +, \cdot)$ *is a field.*

Now we will observe the following:

Theorem 1.7.7. *The system* $(\mathbb{R}, +, \cdot, <)$ *is a complete ordered field.*

Proof. Let $\alpha = (L_1, U_1)$, $\beta = (L_2, U_2)$, and $\gamma = (L_3, U_3)$ be real numbers. We have already observed that one and only one of the following holds:

$$\alpha < \beta \quad \text{or} \quad \alpha = \beta \quad \text{or} \quad \beta < \alpha.$$

We can easily observe that if $\alpha < \beta$ and $\gamma > \mathbf{0}$, then $\alpha\gamma < \beta\gamma$.

Let $\alpha < \beta$. We will show that $\alpha + \gamma < \beta + \gamma$ for every real number γ. Suppose that $\alpha + \gamma = (L, U)$ and $\beta + \gamma = (L', U')$. We have to show that L is properly contained in L'.

By the assumption, L_1 is properly contained in L_2. Therefore L is contained in L'. Note that there is $x \in L_2$ such that $x \notin L_1$. Therefore $x \in U_1$. Since L_2 has no largest element, there is $y > x$ in L_2. By Proposition 1.7.5 there are $z \in L_3$ and $w \in U_3$ such that $w - z = y - x$. This shows that $y + z = x + w$. Note that $y + z \in L'$ and $x + w \in U$. Then $y + z = x + w \notin L$. Therefore L is properly contained in L'.

Now let S be a nonempty subset of $\mathbb{R}$ that is bounded above. Let

$$L = \bigcup_{(L_i, U_i) \in S} L_i \quad \text{and} \quad U = \mathbb{Q} \setminus L.$$

We can check that (L, U) is a real number and the least upper bound of S. $\qquad\square$

For $r \in \mathbb{Q}$, we define $L_r = \{x \in \mathbb{Q} \mid x < r\}$ and $U_r = \{x \in \mathbb{Q} \mid x \geq r\}$. We can check that (L_r, U_r) is a Dedekind cut in which U_r has the smallest element. This gives the embedding $f : \mathbb{Q} \to \mathbb{R}$ defined by $f(r) = (L_r, U_r)$.

Let (L, U) be a Dedekind cut in which U has the smallest element, say $r \in \mathbb{Q}$. Let $x \in L$. Then $x < r$. If $y \in \mathbb{Q}$ is such that $y < r$, then $y \in L$; otherwise, $y \in U$, which will show that $r \leq y$. Hence $L = L_r$ and $U = U_r$. Therefore the Dedekind cut (L, U) in which U has the smallest element determines a rational number. In view of embedding, we say that every rational is a real number. We have seen the existence of a Dedekind cut (L, U) in which U does not have the smallest element. The real number that is not a rational is called an irrational number. Now we observe the following important property of $\mathbb{R}$.

Theorem 1.7.8 (Rational density theorem). *There is a rational between two distinct real numbers.*

Proof. Let $\alpha = (L_1, U_1) < (L_2, U_2) = \beta$. Then L_1 is properly contained in L_2. This implies that there is $a \in L_2$ such that $a \notin L_1$. Now there is an element $b > a$ in L_2. Let $L = \{x \in \mathbb{Q} \mid x < b\}$ and $U = \{x \in \mathbb{Q} \mid x \geq b\}$. Note that $b \in L_2 \setminus L$ and $a \in L \setminus L_1$. This shows that

$$(L_1, U_1) < (L, U) < (L_2, U_2). \qquad\square$$

Now we will observe that the real number system is unique in some sense.

Theorem 1.7.9. *Let $(\mathbb{F}, +, \cdot, <)$ be a complete ordered field. Then there is a bijective map $\phi : \mathbb{F} \to \mathbb{R}$ such that*

(i) $\phi(x + y) = \phi(x) + \phi(y)$,

(ii) $\phi(xy) = \phi(x)\phi(y)$,

(iii) *if $x < y$, then $\phi(x) < \phi(y)$.*

Proof. Let $f : \mathbb{Q} \to \mathbb{F}$ be an embedding (see Exercise 1.23). Given $z \in \mathbb{F}$, we define $L_z = \{x \in \mathbb{Q} \mid f(x) < z\}$ and $U_z = \mathbb{Q} \setminus L_z$. We can check that the map $\phi : \mathbb{F} \to \mathbb{R}$ defined by $\phi(z) = (L_z, U_z)$ is the required one. $\qquad\square$

We now provide the following number system.

Clearly, the equation $X^2 + 1 = 0$ has no solution in the set of real numbers. Let i denote the symbol such that $i^2 + 1 = 0$. Consider the set

$$\mathbb{C} = \{a + bi \mid a, b \in \mathbb{R}\}.$$

The element $z = a + bi$ is called a complex number. The element a is called the real part of z, and we denote it by $a = \mathrm{Re}(z)$. The element b is called the imaginary part of z, and we denote it by $b = \mathrm{Im}(z)$. Two complex numbers $a + bi$ and $c + di$ are called equal if their respective real and imaginary parts are equal. We define the addition and multiplication in the set of complex numbers as follows:

$$(a + bi) + (c + di) = (a + c) + (b + d)i,$$

$$(a + bi)(c + di) = (ac - bd) + (ad + bc)i.$$

We can check that $(\mathbb{C}, +, \cdot)$ is a field and there is no order on $\mathbb{C}$ that extends the order of $\mathbb{R}$. The complex number $\bar{z} = a - bi$ is called the conjugate of $z = a + bi$, and the nonnegative real number $|z| = \sqrt{a^2 + b^2}$ is called the modulus of $z = a + bi$. We can check that $|z|^2 = z\bar{z}$.

Exercises

1.1. First of all, complete whatever is left for you as an exercise.

1.2. Construct the truth table of $((p \Rightarrow q) \wedge q) \Rightarrow r$ and $(p \Rightarrow q) \wedge (q \Rightarrow p)$. Observe whether the bracket arrangement has any role or not.

1.3. If p and q are mathematical statements, then show that the following are tautologically equivalent:

(i) $p \Leftrightarrow q$ and $(p \wedge q) \vee (\neg p \wedge \neg q)$,

(ii) $(p \wedge q) \Rightarrow r$ and $p \Rightarrow (q \Rightarrow r)$.

1.4. Consider the following statements in the following box.

> 1. $\sqrt{2}$ is a rational number.
> 2. $1 + 1 = 0$.
> 3. The line number 3 in this box is false.
> 4. There is something which cannot be proved.
> 5. If a problem is not solvable, then it is not a problem.

Decide the truth or falsity of the statement number 3 in the above box.

1.5. Let A, B, and C be sets. Then show that
   (i) $A \times \emptyset = \emptyset \times A = \emptyset$,
   (ii) $A \times B = B \times A \Leftrightarrow A = B$ or $A = \emptyset$ or $B = \emptyset$,
   (iii) $A \times B = \emptyset \Leftrightarrow A = \emptyset$ or $B = \emptyset$,
   (iv) $A \times (B \cup C) = (A \times B) \cup (A \times C)$,
   (v) $A \times (B \cap C) = (A \times B) \cap (A \times C)$,
   (vi) $(A \setminus B) \times C = (A \times C) \setminus (B \times C)$.

1.6. Show that
   (i) $A \cap \emptyset = \emptyset$,
   (ii) $A \cup \emptyset = A$,
   (iii) $A \setminus \emptyset = A$,
   (iv) $\emptyset \setminus A = \emptyset$.

1.7. Show that
   (i) $\mathcal{P}(A \cap B) = \mathcal{P}(A) \cap \mathcal{P}(B)$;
   (ii) $\mathcal{P}(A) \cup \mathcal{P}(B) \subseteq \mathcal{P}(A \cup B)$; provide an example showing that equality need not hold.

1.8. Let R, S, and T be relations on a set X. Then show that
   (i) $(S \cup T) \circ R = (S \circ R) \cup (T \circ R)$,
   (ii) $(S \cap T) \circ R \subseteq (S \circ R) \cap (T \circ R)$,
   (iii) $S \circ (T \cup R) = (S \circ T) \cup (S \circ R)$,
   (iv) $S \circ (T \cap R) \subseteq (S \circ T) \cap (S \circ R)$.

1.9. Let R be a relation on a set X. Show that $\Delta_X \circ R = R = R \circ \Delta_X$. Provide an example of relations R, S, and T on a set X such that $S \circ R = R$ or $R \circ T = R$ but it is not necessary that $S = \Delta_X$ or $T = \Delta_X$.

1.10. Let R be a relation from a set X to a set Y, and let S be a relation from Y to a set Z. Then show that
   (i) $(R^{-1})^{-1} = R$,
   (ii) $(S \circ R)^{-1} = R^{-1} \circ S^{-1}$.

1.11. Let R and S be equivalence relations on X. Show that $R \circ S$ is an equivalence relation if and only if $R \circ S = S \circ R$.

1.12. Let $f : X \to Y$ and $g : Y \to Z$ be maps. Then show that
   (i) if f and g are injective, then $g \circ f$ is injective;
   (ii) if f and g are surjective, then $g \circ f$ is surjective;
   (iii) if f and g are bijective, then $g \circ f$ is bijective;
   (iv) if $g \circ f$ is injective, then f is injective;

(v)   if $g \circ f$ is surjective, then g is surjective;

(vi)   if $g \circ f$ is bijective, then g is surjective, and f is injective.

1.13.   Let $f : X \to Y$ and $g : Y \to Z$ be bijective maps. Then show that

(i) $(f^{-1})^{-1} = f$;

(ii) $f^{-1} : Y \to X$ is a bijective map;

(iii) $f^{-1} \circ f = I_X$, and $f \circ f^{-1} = I_Y$;

(iv) $(g \circ f)^{-1} = f^{-1} \circ g^{-1}$.

1.14.   Show that there is no surjective map from a set to its power set.

1.15.   Let X and Y be sets. Then Y^X denotes the set of all maps from X to Y. Show that there is a bijective map from 2^X to the power set $\mathcal{P}(X)$, where $2 = \{0, 1\}$.

1.16.   Let $f : X \to Y$ be a map. Then show that

(i)   for each $A \subseteq X, A \subseteq f^{-1}(f(A))$;

(ii) $A = f^{-1}(f(A))$ for all $A \subseteq X$ if and only if f is injective;

(iii)   for all $A \subseteq X$ and $B \subseteq Y, f(f^{-1}(B) \cap A) = B \cap f(A)$;

(iv)   for each $B \subseteq Y, f(f^{-1}(B)) \subseteq B$;

(v) $B = f(f^{-1}(B))$ for all $B \subseteq Y$ if and only if f is surjective.

1.17.   Let I and J be indexing sets. Let $\{A_\alpha \mid \alpha \in I\}$ and $\{B_\alpha \mid \alpha \in J\}$ be families of sets. Then show that

(i)   (De Morgan's law) for a set $X, X - (\bigcup_{\alpha \in I} A_\alpha) = \bigcap_{\alpha \in I}(X - A_\alpha)$;

(ii)   (De Morgan's law) for a set $X, X - (\bigcap_{\alpha \in I} A_\alpha) = \bigcup_{\alpha \in I}(X - A_\alpha)$;

(iii)   if $I = \bigcup_{\beta \in J} I_\beta$, then

(a) $\bigcup\{A_\alpha \mid \alpha \in I\} = \bigcup_{\beta \in J}(\bigcup\{A_\alpha \mid \alpha \in I_\beta\})$,

(b) $\bigcap\{A_\alpha \mid \alpha \in I\} = \bigcap_{\beta \in J}(\bigcap\{A_\alpha \mid \alpha \in I_\beta\})$;

(iv) $(\bigcup\{A_\alpha \mid \alpha \in I\}) \cap (\bigcup\{B_\beta \mid \beta \in J\}) = \cup(A_\alpha \cap B_\beta \mid \alpha \in I \text{ and } \beta \in J)$;

(v) $(\bigcap\{A_\alpha \mid \alpha \in I\}) \cup (\bigcap\{B_\beta \mid \beta \in J\}) = \bigcap(A_\alpha \cup B_\beta \mid \alpha \in I \text{ and } \beta \in J)$.

1.18.   A set X is called a finite set if there is a bijection between X and some $n \in \mathbb{N}_0$; otherwise, X is called an infinite set. Show that if X is finite, then

(i)   every injective map $f : X \to X$ is surjective;

(ii)   every surjective map $f : X \to X$ is injective.

1.19.   (Pigeonhole principle) Let X and Y be finite sets containing m and n elements, respectively, where $m, n \in \mathbb{N}$ and $n < m$. Then there is no injective map from X to Y.

1.20.   (Division algorithm) Let $x, y \in \mathbb{Z}$ and $y > 0$. Show that there exists a unique pair $(q, r) \in \mathbb{Z} \times \mathbb{Z}$ such that $x = yq + r$ with $0 \leq r < y$.

1.21.   Show that

(i)   any nonempty subset of $\mathbb{Z}$ that is bounded above in $\mathbb{Z}$ has the largest element;

(ii)   any nonempty subset of $\mathbb{Z}$ that is bounded below in $\mathbb{Z}$ has the least element.

1.22.   Let $\mathbb{F}$ be an ordered field. Show that

(i) $x^2 > 0$ for every $x \in \mathbb{F} \setminus \{0\}$;

(ii) $1 > 0$;

(iii)   if $xy > 0$ and $y > 0$, then $x > 0$;

(iv)   if $x > 0$, then $x^{-1} = \frac{1}{x} > 0$;

(v)   if $x > 0, y > 0$, and $n \in \mathbb{N}$, then $x < y \Leftrightarrow x^n < y^n$.

1.23. Let $\mathbb{F}$ be an ordered field with multiplicative identity 1. For $m \in \mathbb{Z}$, define

$$
m_{\mathbb{F}} = \begin{cases} 1 + 1 + \cdots + 1 \,(m \text{ times}) & \text{if } m > 0, \\ 0 & \text{if } m = 0, \\ (-1) + (-1) + \cdots + (-1)(-m \text{ times}) & \text{if } m < 0. \end{cases}
$$

Show that the map $f : \mathbb{Q} \to \mathbb{F}$ defined by $f(\frac{m}{n}) = m_{\mathbb{F}} n_{\mathbb{F}}^{-1}$ is an embedding of $\mathbb{Q}$ in $\mathbb{F}$. Hence deduce that no finite field can be ordered.

1.24. Let $\mathbb{F}$ be an ordered field. Define the map $|\cdot| : \mathbb{F} \to \mathbb{R}$ by $|x| = x$ if $x \geq 0$ and $|x| = -x$ if $x < 0$. Show that

(i) $|x| \geq 0$ and $|x| = 0$ if and only if $x = 0$;

(ii) $|x| = |-x|$;

(iii) $|xy| = |x||y|$;

(iv) $|x + y| \leq |x| + |y|$.

1.25. Let (L, U) be a Dedekind cut, and let $n \in \mathbb{N}$. Show that there are $x \in L$ and $y \in U$ such that $y - x < \frac{1}{n}$.

1.26. (Irrational density theorem) Show that there is an irrational between two distinct real numbers.

1.27. (Archimedean property) Let $x, y \in \mathbb{R}$ with $x > 0$. Show that there is $n \in \mathbb{N}$ such that $nx > y$.

1.28. Let A and B partition the set of real numbers $\mathbb{R}$ so that if $a \in A$ and $b \in B$, then $a < b$. Show that there is a unique real number c such that $a \leq c$ and $c \leq b$ for all $a \in A$ and $b \in B$.

2 Metric spaces

In this chapter, we define the notion of a distance or metric on a set. This will be a foundation or motivation of the way we will define topology latter.

2.1 Metric spaces: examples

Consider two points $x = (x_1, x_2)$ and $y = (y_1, y_2)$ in $\mathbb{R}^2$. We know that the distance or metric $d(x, y)$ between x and y is given as

$$d(x, y) = \sqrt{(x_1 - y_1)^2 + (x_2 - y_2)^2}.$$

This defines a map $d : \mathbb{R}^2 \times \mathbb{R}^2 \to \mathbb{R}$ given by $(x, y) \mapsto d(x, y)$. It can be verified that the map d satisfies the following properties:

(i) $d(x, y) \geq 0$,
(ii) $d(x, y) = 0$ if and only if $x = y$,
(iii) $d(x, y) = d(y, x)$,
(iv) $d(x, y) \leq d(x, z) + d(z, y)$,

where $x, y, z \in \mathbb{R}^2$. Property (iii) is called the symmetry of d, and (iv) is called the triangle inequality. The triangle inequality simply says that the sum of the lengths of two sides of a triangle is greater than or equal to the length of the third side. These properties give us a sense of distance or metric on a set.

Definition 2.1.1. Let X be a nonempty set. Then a map $d : X \times X \to \mathbb{R}$ is called a metric on X if for all $x, y, z \in X$, the following conditions hold:

(i) $d(x, y) \geq 0$,
(ii) $d(x, y) = 0$ if and only if $x = y$,
(iii) $d(x, y) = d(y, x)$,
(iv) $d(x, y) \leq d(x, z) + d(z, y)$.

The pair (X, d), where X is a nonempty set, and d a metric on X, is called a metric space. Thus a metric d on a nonempty set is a nonnegative real-valued map that is symmetric, satisfies the triangle inequality, and is zero only on the diagonal of X.

Let (X, d) be a metric space, and let $x, y \in X$. Then $d(x, y)$ is called the distance between x and y. When the metric d on a nonempty set X is given, we will only say that X is a metric space or space. An element x in a metric space X is called a point in X. By a set A in a metric space X we will mean that A is a subset of X.

Example 2.1.2. Consider the set $\mathbb{R}$ of real numbers. Define $d : \mathbb{R} \times \mathbb{R} \to \mathbb{R}$ by $d(x, y) = |x - y|$. Then d is a metric on $\mathbb{R}$. The set $\mathbb{R}$ with this metric is called the real line.

https://doi.org/10.1515/9783111636085-002

Example 2.1.3. Consider the set $\mathbb{R}^n$ ($n \in \mathbb{N}$). Let $p \in \mathbb{R}$ be such that $p \geq 1$, and let $x = (x_1, \ldots, x_n)$ and $y = (y_1, \ldots, y_n)$ be in $\mathbb{R}^n$. Define $d_p : \mathbb{R}^n \times \mathbb{R}^n \to \mathbb{R}$ by

$$d_p(x, y) = \left(\sum_{i=1}^{n} |x_i - y_i|^p \right)^{\frac{1}{p}}.$$

Then d_p is a metric on $\mathbb{R}^n$. We will prove the triangle inequality. First, note that for $p \geq 1$, the map $f : \mathbb{R} \to \mathbb{R}$ defined by $f(t) = |t|^p$ is convex (for $p = 1$, it is the consequence of the triangle inequality on $\mathbb{R}$, and for $p > 1$, the function f is twice differentiable and the derivative $f''(t) \geq 0$). This implies that for $0 \leq \alpha \leq 1$,

$$|\alpha u + (1 - \alpha)v|^p \leq \alpha |u|^p + (1 - \alpha)|v|^p,$$

where $u, v \in \mathbb{R}$. Therefore, for $a = (a_1, \ldots, a_n)$ and $b = (b_1, \ldots, b_n)$, we have

$$\sum_{i=1}^{n} |\alpha a_i + (1 - \alpha)b_i|^p \leq \sum_{i=1}^{n} (\alpha |a_i|^p + (1 - \alpha)|b_i|^p)$$

$$= \alpha \sum_{i=1}^{n} |a_i|^p + (1 - \alpha) \sum_{i=1}^{n} |b_i|^p. \tag{2.1}$$

Let $A = \left(\sum_{i=1}^{n} |a_i|^p \right)^{\frac{1}{p}}$ and $B = \left(\sum_{i=1}^{n} |b_i|^p \right)^{\frac{1}{p}}$. Then, replacing a_i by $\frac{a_i}{A}$ and b_i by $\frac{b_i}{B}$ in equation (2.1), we get

$$\sum_{i=1}^{n} \left| \alpha \frac{a_i}{A} + (1 - \alpha)\frac{b_i}{B} \right|^p \leq \sum_{i=1}^{n} \left(\alpha \frac{|a_i|^p}{A^p} + (1 - \alpha)\frac{|b_i|^p}{B^p} \right)$$

$$= 1. \tag{2.2}$$

Taking $\alpha = \frac{A}{A+B}$ in (2.2), we get

$$\left(\sum_{i=1}^{n} |a_i + b_i|^p \right)^{\frac{1}{p}} \leq \left(\sum_{i=1}^{n} |a_i|^p \right)^{\frac{1}{p}} + \left(\sum_{i=1}^{n} |b_i|^p \right)^{\frac{1}{p}}. \tag{2.3}$$

Taking $a_i = x_i - y_i$ and $b_i = y_i - z_i$ in (2.3), we get the required triangle inequality.

Example 2.1.4. Define $d_\infty : \mathbb{R}^n \times \mathbb{R}^n \to \mathbb{R}$ by $d_\infty(x, y) = \max\{|x_i - y_i| \mid 1 \leq i \leq n\}$ for $x = (x_1, \ldots, x_n)$ and $y = (y_1, \ldots, y_n)$ in $\mathbb{R}^n$. Then d_∞ is a metric on $\mathbb{R}^n$.

Note 2.1.5. For $p = 2$ in Example 2.1.3, the metric d_2 is called the Euclidean or usual metric on $\mathbb{R}^n$, and the metric space $\mathbb{R}^n$ is called the Euclidean space. Whenever we use $\mathbb{R}^n$ without mentioning the metric on it, we will mean that it is with the usual metric.

Example 2.1.6. Let X be a nonempty set. Define $d : X \times X \to \mathbb{R}$ by

$$d(x,y) = \begin{cases} 0 & \text{if } x = y, \\ 1 & \text{if } x \neq y. \end{cases}$$

Then d is a metric on X. This metric is called the discrete metric on X, and the space (X, d) is called the discrete metric space.

Note 2.1.7. Let u be any positive real number. Then

$$d(x,y) = \begin{cases} 0 & \text{if } x = y, \\ u & \text{if } x \neq y, \end{cases}$$

is also a metric on X. We can also call this metric the discrete metric on X.

Example 2.1.8. Let (X, d) be a metric space. Then the map $\overline{d} : X \times X \to \mathbb{R}$ defined by

$$\overline{d}(x,y) = \frac{d(x,y)}{1 + d(x,y)}$$

is a metric on X. Conditions (i), (ii), and (iii) of Definition 2.1.1 are easy to verify. We will verify the triangle inequality. Let $x, y, z \in X$. Then

$$\begin{aligned} \overline{d}(x,z) + \overline{d}(z,y) &= \frac{d(x,z)}{1 + d(x,z)} + \frac{d(z,y)}{1 + d(z,y)} \\ &\geq \frac{d(x,z)}{1 + d(x,z) + d(z,y)} + \frac{d(z,y)}{1 + d(x,z) + d(z,y)} \\ &= \frac{d(x,z) + d(z,y)}{1 + d(x,z) + d(z,y)} \\ &= \frac{1}{1 + \frac{1}{d(x,z)+d(z,y)}} \\ &\geq \frac{1}{1 + \frac{1}{d(x,y)}} \\ &= \overline{d}(x,y). \end{aligned}$$

Note that $\overline{d}(x,y) < 1$ for all $x, y \in X$.

Example 2.1.9. Let (X, d) be a metric space. Define $\overline{d} : X \times X \to \mathbb{R}$ by $\overline{d}(x,y) = \min\{1, d(x,y)\}$. Then $\overline{d}$ is a metric on X. For the triangle inequality, let $d(x,z) \leq 1$ and $d(z,y) \leq 1$. Then

$$\overline{d}(x,y) \leq d(x,y) \leq d(x,z) + d(z,y) = \overline{d}(x,z) + \overline{d}(z,y).$$

Now let $d(x,z) > 1$. Then

$$\bar{d}(x,z) \le 1 \le 1 + \bar{d}(z,y) = \bar{d}(x,z) + \bar{d}(z,y).$$

We can similarly check the triangle inequality for the case $d(z,y) > 1$. Note that $\bar{d}(x,y) \le 1$ for all $x, y \in X$. This metric is called the standard bounded metric corresponding to d.

Example 2.1.10. Let X be a nonempty set. Let $\mathcal{F}$ be the collection of all finite subsets of X. For subsets A and B of X, let $A \triangle B = (A \setminus B) \cup (B \setminus A)$. For $A \in \mathcal{F}$, let $|A|$ denote the number of elements in A. Define $d : \mathcal{F} \times \mathcal{F} \to \mathbb{R}$ by

$$d(A,B) = |A \triangle B|.$$

Then d is a metric on $\mathcal{F}$. Conditions (i) and (iii) of Definition 2.1.1 are easy to verify. For (ii), let $A = B$. Then $d(A,B) = 0$. Conversely, $d(A,B) = 0$ implies that $A \setminus B = \emptyset$ and $B \setminus A = \emptyset$. Therefore $A = B$. This proves (ii).

It is easy to verify that $A \triangle B \subseteq (A \triangle C) \cup (C \triangle B)$. Now

$$
\begin{aligned}
d(A,C) + d(C,B) &= |A \triangle C| + |C \triangle B| \\
&= \left|(A \triangle C) \cup (C \triangle B)\right| + \left|(A \triangle C) \cap (C \triangle B)\right| \\
&\ge \left|(A \triangle C) \cup (C \triangle B)\right| \\
&\ge |A \triangle B| = d(A,B).
\end{aligned}
$$

This proves (iv).

Example 2.1.11. Let p be a prime. Define $d_p : \mathbb{Z} \times \mathbb{Z} \to \mathbb{R}$ by

$$
d_p(x,y) = \begin{cases} 0 & \text{if } x = y, \\ \dfrac{1}{p^{\max\{n \in \mathbb{N} | p^n \text{ divides } (x-y)\}}} & \text{if } x \ne y. \end{cases}
$$

Then d_p is a metric on $\mathbb{Z}$. For the triangle inequality, let $x, y, z \in \mathbb{Z}$ be such that $x \ne y$, $y \ne z$, and $x \ne z$. Let

$$\theta(x,y) = \max\{n \in \mathbb{N} \mid p^n \text{ divides } (x-y)\}.$$

We can assume that $\theta(x,z) \le \theta(z,y)$. This implies that $p^{\theta(x,z)}$ divides $x - z$ and $z - y$. Therefore $p^{\theta(x,z)}$ divides $x - y$. Hence $\theta(x,z) \le \theta(x,y)$. This shows that

$$d_p(x,y) \le d_p(x,z) = \max\{d_p(x,z), d_p(z,y)\}.$$

Thus $d_p(x,y) \le d_p(x,z) + d_p(z,y)$. The other cases for the triangle inequality can be easily proved.

Example 2.1.12. Let p be a prime. For any nonzero $x \in \mathbb{Q}$, there exists a unique $n \in \mathbb{Z}$ such that $x = p^n \frac{a}{b}$, where p does not divide a and b. Let $v_p(x) = n$. Define $d_p : \mathbb{Q} \times \mathbb{Q} \to \mathbb{R}$ by

$$d_p(x,y) = \begin{cases} 0 & \text{if } x = y, \\ \frac{1}{p^{v_p(x-y)}} & \text{if } x \neq y. \end{cases}$$

Then d_p is a metric on $\mathbb{Q}$. To prove the triangle inequality, we can prove that $d_p(x,y) \leq \max\{d_p(x,z), d_p(z,y)\}$. Note that the metric d_p in Example 2.1.11 is the restriction of this metric on $\mathbb{Z}$.

Remark 2.1.13. The metric space (X, d) in which the condition

$$d(x,y) \leq \max\{d(x,z), d(z,y)\}$$

is satisfied for all $x, y, z \in X$ is called an ultrametric space. In an ultrametric space, all the triangles are isosceles. For this, we have to show that either of two distances among $x, y, z \in X$ must be equal. On the contrary, suppose that all the distances among $x, y, z \in X$ are different. We can assume that $d(x,y) < d(x,z) < d(y,z)$, but this will contradict the condition $d(x,y) \leq \max\{d(x,z), d(z,y)\}$.

Example 2.1.14. Let (X, d) be a metric space, and let A be a nonempty subset of X. Define $d_A : X \times X \to \mathbb{R}$ by $d_A(x,y) = d(x,y)$. Then (A, d_A) is a metric space. This space is called the metric subspace or subspace of X. The metric space in Example 2.1.11 is a metric subspace of the metric space in Example 2.1.12.

Example 2.1.15. Let (Y, d) be a metric space, and X let be a set. Let $f : X \to Y$ be a bijective map. We define $d' : X \times X \to \mathbb{R}$ by $d'(x,y) = d(f(x), f(y))$. Then d' is a metric on X.

Example 2.1.16. Consider the real line $\mathbb{R}$. Let $-\infty$ and ∞ be two different symbols not in $\mathbb{R}$. Let $\tilde{\mathbb{R}} = \mathbb{R} \cup \{-\infty, \infty\}$. We extend the order of $\mathbb{R}$ to an order of $\tilde{\mathbb{R}}$ by putting $-\infty < \infty$ and $-\infty < x < \infty$ for all $x \in \mathbb{R}$. We can check that $\tilde{\mathbb{R}}$ is a totally ordered set with respect to this order. We denote the set $\tilde{\mathbb{R}}$ by $[-\infty, \infty]$ and $\mathbb{R}$ by $(-\infty, \infty)$. Also, we denote the sets $\{x \in \mathbb{R} \mid x > a\}$ and $\{x \in \mathbb{R} \mid x < a\}$ by (a, ∞) and $(-\infty, a)$, respectively. The $(a, \infty]$ and $[-\infty, a)$ have their meanings accordingly.

Define the map $f : \mathbb{R} \to (-1, 1)$ by $f(x) = \frac{x}{1+|x|}$. We can check that f is bijective and strictly increasing. This shows that for each $a \in \mathbb{R}$, we have $f(a, \infty) = (f(a), 1)$ and $f(-\infty, a) = (-1, f(a))$. By elementary analysis we can observe that $\lim_{x \to \infty} f(x) = 1$ and $\lim_{x \to -\infty} f(x) = -1$. This prompts us to extend the map f to the map $\tilde{f} : \tilde{\mathbb{R}} \to [-1, 1]$ by putting $f(-\infty) = -1$ and $f(\infty) = 1$. Clearly, $\tilde{f}$ is a bijective map. Note that $[-1, 1]$ is a metric subspace of $\mathbb{R}$. Therefore we can induce a metric $\tilde{d}$ on $\tilde{\mathbb{R}}$ defined through $\tilde{f}$ by $\tilde{d}(x,y) = |\tilde{f}(x) - \tilde{f}(y)|$. This metric space is called the extended real line.

Definition 2.1.17. Let (X, d) be a metric space, and let r be a positive real number. Then an open ball $B_d(x, r)$ centered at $x \in X$ and radius r is the subset of X consisting of all the points that are at distance less than r from x. In other words,

$$B_d(x, r) = \{y \in X \mid d(x,y) < r\}.$$

When the metric d is given, we denote the open ball by $B(x, r)$. The open ball $B(x, r)$ in a metric space X is nonempty since $x \in B(x, r)$. If $B(x, r)$ is open ball centered at x, then we also say that it is an open ball around x.

Definition 2.1.18. Let (X, d) be a metric space, and let r be a positive real number. Then a closed ball $D(x, r)$ centered at $x \in X$ and radius r is the subset of X consisting of all the points that are at distance less than or equal to r from x. In other words,

$$D(x, r) = \{y \in X \mid d(x, y) \leq r\}.$$

Example 2.1.19. Consider $\mathbb{R}$ with the usual metric. Then $B(x, r) = (x - r, x + r)$.

Example 2.1.20. Consider the extended real line $\tilde{\mathbb{R}}$. Let us find the open ball around ∞. Let $0 < r < 1$. We claim that $B(\infty, r) = (\frac{1}{r} - 1, \infty]$. If $x \in (\frac{1}{r} - 1, \infty]$, then $x > 0$ for $\frac{1}{r} - 1 > 0$. If $x \in B(\infty, r)$, then $x > 0$ for $r < 1$. Let $x > 0$. Then

$$\tilde{d}(x, \infty) = |\tilde{f}(x) - \tilde{f}(\infty)| = \left| \frac{x}{1+x} - 1 \right| = \frac{1}{1+x}.$$

This shows that $x \in B(\infty, r) \Leftrightarrow x \in (\frac{1}{r} - 1, \infty]$. Similarly, we can observe that for $0 < r < 1$, both the sets $B(-\infty, r)$ and $[-\infty, 1 - \frac{1}{r})$ contain the numbers $x < 0$. Then, for $x < 0$, $\tilde{d}(x, -\infty) = \frac{1}{1-x}$. This shows that $B(-\infty, r) = [-\infty, 1 - \frac{1}{r})$.

We should not get confused with the terminology of the ball or the center. A ball in a metric space may not look like a ball or center of our geometric imagination. For example, consider the discrete metric space X. Then

$$B(x, r) = \begin{cases} \{x\} & \text{if } r \leq 1, \\ X & \text{if } r > 1. \end{cases}$$

Also, in an ultrametric space X, each point of an open ball is a center. To see this, let $y \in B(x, r)$, that is, $d(x, y) < r$, and $z \in B(y, r)$, that is, $d(y, z) < r$. Then

$$d(x, z) \leq \max\{d(x, y), d(y, z)\} < r.$$

This implies that $B(y, r) \subseteq B(x, r)$. We can similarly show that $B(x, r) \subseteq B(y, r)$. Thus $B(x, r) = B(y, r)$. However, note that if B_1 and B_2 are two open balls with the same center in a metric space X, then either $B_1 \subseteq B_2$ or $B_2 \subseteq B_1$.

Example 2.1.21. Consider $\mathbb{R}^2$ with the metric $d_1((x_1, x_2), (y_1, y_2)) = |x_1 - y_1| + |x_2 - y_2|$. Then the open ball $B((0, 0), 1)$ consists of all the points inside the quadrilateral bounded by lines $x + y < 1$, $-x + y < 1$, $x - y < 1$, and $-x - y < 1$ (see Figure 2.1).

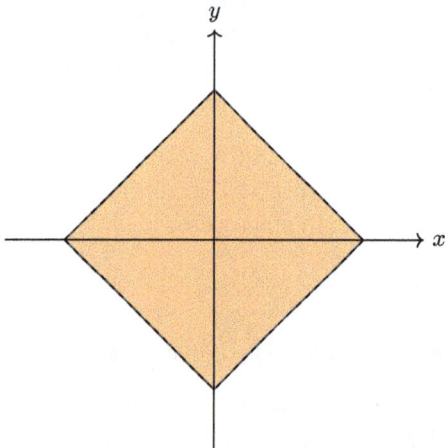

Figure 2.1: Open ball in $(\mathbb{R}^2, d_1)$.

Example 2.1.22. Consider $\mathbb{R}^3$ with the metric

$$d_1((x_1, x_2, x_3), (y_1, y_2, y_3)) = |x_1 - y_1| + |x_2 - y_2| + |x_3 - y_3|.$$

Then the open ball $B((0, 0, 0), 1)$ consists of all the points inside the octahedron bounded by the planes $|x| + |y| + |z| < 1$ (see Figure 2.2).

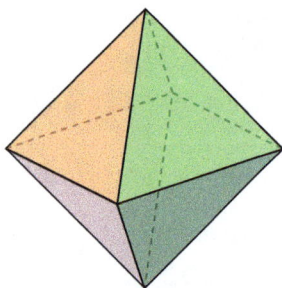

Figure 2.2: Open ball in $(\mathbb{R}^3, d_1)$.

Proposition 2.1.23. *Let X be a metric space. Let $x, y \in X$ be such that $x \neq y$. Then there exist disjoint open balls centered at x and y, respectively.*

Proof. First, note that $r = d(x, y) > 0$. Then we will show that $B(x, \frac{r}{2}) \cap B(y, \frac{r}{2}) = \emptyset$. On the contrary, suppose that $z \in B(x, \frac{r}{2}) \cap B(y, \frac{r}{2})$. Then $d(x, z) < \frac{r}{2}$ and $d(y, z) < \frac{r}{2}$. Now by the triangle inequality we have

$$r = d(x,y) \le d(x,z) + d(y,z) < \frac{r}{2} + \frac{r}{2} = r,$$

a contradiction. □

Definition 2.1.24. A set A in a metric space X is called a bounded set if the set $\{d(x,y) \mid x,y \in A\}$ is a bounded set in $\mathbb{R}$.

Note that a set A in a metric space X is bounded if and only if A is contained in some open ball. A metric space X is called bounded if X itself is a bounded set. In this case the metric on X is called the bounded metric. The diameter diam(A) of a nonempty bounded set A in X is defined as

$$\text{diam}(A) = \sup\{d(x,y) \mid x,y \in A\},$$

where $\sup T$ denotes the supremum or least upper bound of a subset T of $\mathbb{R}$. We define the diameter of the empty set as 0.

Example 2.1.25. The sets $(0,1)$ and $\{\frac{1}{n} \mid n \in \mathbb{N}\}$ are bounded sets in $\mathbb{R}$, whereas $\mathbb{N}$ and $\{x \in \mathbb{R} \mid x > 0\}$ are not bounded sets in $\mathbb{R}$. Also, note that $\text{diam}(0,1) = \text{diam}\{\frac{1}{n} \mid n \in \mathbb{N}\} = 1$.

Example 2.1.26. The discrete metric space is bounded. The metric spaces defined in Examples 2.1.8 and 2.1.9 are bounded metric spaces.

2.2 Normed linear spaces

In this section, we define the norm on a vector space. A norm on a vector space induces a metric structure on it. We will observe that this metric depends on the algebraic structure of the vector space.

Definition 2.2.1. Let V be a vector space over a field $\mathbb{F}$, where $\mathbb{F}$ is either $\mathbb{R}$ or $\mathbb{C}$. Then a map $\| \cdot \| : V \to \mathbb{R}$ is called a norm if the following conditions are satisfied:
(i) $\|x\| \ge 0$,
(ii) $\|x\| = 0$ if and only if $x = 0$,
(iii) $\|\alpha x\| = |\alpha| \|x\|$ for all $x \in V$ and $\alpha \in \mathbb{F}$,
(iv) $\|x + y\| \le \|x\| + \|y\|$ for all $x,y \in V$.

If $\| \cdot \|$ is a norm on a vector space V, then we say that the pair $(V, \| \cdot \|)$ is a normed linear space or normed space. Let $(V, \| \cdot \|)$ be a normed space. Define a map $d : V \times V \to \mathbb{R}$ by $d(x,y) = \|x - y\|$. We can check that d is a metric on V. This is called the metric induced by a norm. From condition (iii) of Definition 2.2.1 we observe that the metric induced by a norm cannot be bounded. This shows that a bounded metric cannot be induced by a norm. When a norm $\| \cdot \|$ on V is given, we will only say that V is a normed space. We will also call condition (iv) of Definition 2.2.1 the triangle inequality. Throughout the book the field $\mathbb{F}$ will denote either $\mathbb{R}$ or $\mathbb{C}$; otherwise, it will be stated explicitly.

Let V be a normed space, let $x \in V$, and let r be a positive real number. Let $A \subseteq V$. Then by the sets $x + A$ and rA we mean the sets $\{x + y \mid y \in A\}$ and $\{ry \mid y \in A\}$. Let d be the metric on V induced by the norm. Now

$$
\begin{aligned}
B(x,r) &= \{y \in V \mid d(x,y) < r\} = \{y \in V \mid \|y - x\| < r\} \\
&= \{x + z \in V \mid \|z\| < r\} = \{x + rw \in V \mid \|w\| < 1\} \\
&= x + rB(0,1).
\end{aligned}
$$

Example 2.2.2. We have shown in Example 2.1.3 that $\mathbb{R}^n$ is a metric space with the metric d_p ($p \geq 1$). In fact, this metric is induced by the norm $\|x\|_p = d_p(x,0)$, where $x \in \mathbb{R}^n$. Now we show that a similar norm can be defined on $\mathbb{C}^n$.

Let $z = x + iy \in \mathbb{C}$. Then $|z| = \sqrt{x^2 + y^2}$ defines a norm on $\mathbb{C}$. For the triangle inequality, observe that for $z_1, z_2 \in \mathbb{C}$, we have

$$
\begin{aligned}
|z_1 + z_2|^2 &= (z_1 + z_2)(\overline{z_1 + z_2}) = z_1\overline{z_1} + z_1\overline{z_2} + z_2\overline{z_1} + z_2\overline{z_2} \\
&= |z_1|^2 + 2\operatorname{Re}(z_1\overline{z_2}) + |z_2|^2 \\
&\leq |z_1|^2 + 2|z_1||\overline{z_2}| + |z_2|^2 \\
&= (|z_1| + |z_2|)^2,
\end{aligned}
$$

where $\bar{z}$ and $\operatorname{Re}(z)$ is the complex conjugate and the real part of a complex number z. This shows that $|z_1 + z_2| \leq |z_1| + |z_2|$. The rest of the conditions are easy to verify.

Let $p > 1$ be a real number. Let $z = (z_1, z_2, \ldots, z_n) \in \mathbb{C}^n$. Define $\|z\|_p = (\sum_{i=1}^n |z_i|^p)^{\frac{1}{p}}$. We will only show the triangle inequality. For the proof, we will adopt the same method as in Example 2.1.3. Recall that $\phi : \mathbb{R}^n \to \mathbb{R}$ is called a convex map if for all $x, y \in \mathbb{R}^n$ and $0 \leq a \leq 1$, we have

$$
\phi(ax + (1 - a)y) \leq a\phi(x) + (1 - a)\phi(y).
$$

From the multivariate calculus we have the following:

Proposition 2.2.3. *A twice differentiable map $f : \mathbb{R}^n \to \mathbb{R}$ is convex if and only if the Hessian matrix of f is positive definite.*

Now identifying $\mathbb{C}$ with $\mathbb{R}^2$, we define the map $f : \mathbb{R}^2 \to \mathbb{R}$ by $f(z) = |z|^p = (x^2 + y^2)^{\frac{p}{2}}$. Note that for $p = 1$, by the triangle inequality as observed above, f is convex. Let $p > 1$. Observe that f is twice differentiable. The Hessian matrix of f is as follows:

$$
\begin{pmatrix}
\frac{\partial^2 f}{\partial x^2} & \frac{\partial^2 f}{\partial x \partial y} \\
\frac{\partial^2 f}{\partial y \partial x} & \frac{\partial^2 f}{\partial y^2}
\end{pmatrix}.
$$

Note that for the given map f,

$$\frac{\partial^2 f}{\partial x^2} \geq 0 \quad \text{and} \quad \frac{\partial^2 f}{\partial x^2}\frac{\partial^2 f}{\partial y^2} - \frac{\partial^2 f}{\partial x \partial y}\frac{\partial^2 f}{\partial y \partial x} \geq 0.$$

This shows that the Hessian matrix is positive definite. Therefore, by Proposition 2.2.3, f is convex. Now by a similar argument as in Example 2.1.3 we have

$$\|z_1 + z_2\|_p \leq \|z_1\|_p + \|z_2\|_p. \tag{2.4}$$

Inequality (2.4) is called the Minkowski inequality.

Example 2.2.4. Define $\|\cdot\|_\infty : \mathbb{F}^n \to \mathbb{R}$ by $\|x\|_\infty = \max\{|x_i| \mid 1 \leq i \leq n\}$ for $x = (x_1, \ldots, x_n) \in \mathbb{F}^n$. Then $\|\cdot\|_\infty$ is a norm on $\mathbb{F}^n$.

Definition 2.2.5. Let X be a nonempty set. A map $f : \mathbb{N} \to X$ is called a sequence in X. The image $f(n)$ of an element $n \in \mathbb{N}$ is called the nth term of the sequence f. If $f(n) = x_n$, then we denote the sequence f by (x_n) or

$$x_1, x_2, \ldots, x_n, \ldots .$$

Example 2.2.6. Let $p > 1$. Consider the set X of sequences $x = (x_n)$ in $\mathbb{F}$ such that $\sum_{n=1}^\infty |x_n|^p < \infty$. Then X is a vector space with pointwise addition and pointwise scalar multiplication. Define $\|\cdot\|_p : X \to \mathbb{R}$ by $\|x\|_p = (\sum_{n=1}^\infty |x_n|^p)^{\frac{1}{p}}$. Then $\|\cdot\|_p$ is a norm on X. We will only prove the triangle inequality. Let $x, y \in X$. Then, for all $n \in \mathbb{N}$, we have

$$\left(\sum_{k=1}^n |x_k + y_k|^p\right)^{\frac{1}{p}} \leq \left(\sum_{k=1}^n |x_k|^p\right)^{\frac{1}{p}} + \left(\sum_{k=1}^n |y_k|^p\right)^{\frac{1}{p}}$$

$$\leq \left(\sum_{k=1}^\infty |x_k|^p\right)^{\frac{1}{p}} + \left(\sum_{k=1}^\infty |y_k|^p\right)^{\frac{1}{p}}.$$

Therefore

$$\left(\sum_{k=1}^\infty |x_k + y_k|^p\right)^{\frac{1}{p}} \leq \left(\sum_{k=1}^\infty |x_k|^p\right)^{\frac{1}{p}} + \left(\sum_{k=1}^\infty |y_k|^p\right)^{\frac{1}{p}}.$$

This normed space is denoted by ℓ_p.

Example 2.2.7. Let X be the set of sequences $x = (x_n)$ in $\mathbb{F}$ such that $\sup\{|x_k| \mid k \in \mathbb{N}\} < \infty$. Define $\|\cdot\|_\infty : X \to \mathbb{R}$ by $\|x\|_\infty = \sup\{|x_k| \mid k \in \mathbb{N}\}$. Then $\|\cdot\|_\infty$ is a norm on X. This normed space is denoted by ℓ_∞.

Example 2.2.8. Let $C_\mathbb{F}[a,b]$ denote the set of all continuous maps from $[a,b]$ to $\mathbb{F}$. Then $C_\mathbb{F}[a,b]$ is a vector space with the following operations:

$$(f+g)(x) = f(x) + g(x), \quad (af)(x) = af(x),$$

where $f, g \in C_{\mathbb{F}}[a,b]$ and $\alpha \in \mathbb{F}$. Define $\| \cdot \| : C_{\mathbb{F}}[a,b] \to \mathbb{R}$ by $\|f\| = \int_a^b |f(t)| dt$. Then $\| \cdot \|$ is a norm on $C_{\mathbb{F}}[a,b]$. We will only prove condition (ii). If $f = 0$, then clearly $\|f\| = 0$. For the converse, suppose $f \neq 0$. Since f is continuous, there exist $c, d \in [a,b]$ with $c < d$ and $\delta > 0$ such that $|f(t)| \geq 0$ for all $t \in [c,d]$. This implies that

$$\|f\| \geq \int_c^d \geq (d-c)\delta > 0.$$

Example 2.2.9. Define $\| \cdot \|_\infty : C_{\mathbb{F}}[a,b] \to \mathbb{R}$ by $\|f\|_\infty = \sup\{|f(t)| \mid t \in [a,b]\}$. Then $\| \cdot \|_\infty$ is a norm on $C_{\mathbb{F}}[a,b]$. We will write $C[a,b]$ instead of $C_{\mathbb{F}}[a,b]$ when the field $\mathbb{F}$ is understood.

Example 2.2.10. We have seen examples of a few normed spaces. In fact, we can construct a norm on any vector space over $\mathbb{F}$. This follows from the fact that every vector space V has a basis $B = \{v_a \mid a \in \mathcal{I}\}$. Each nonzero element x in V can be written as $x = \alpha_{i_1} v_{i_1} + \alpha_{i_2} v_{i_2} + \cdots + \alpha_{i_n} v_{i_n}$, where $\alpha_{i_r} \in \mathbb{F}$ and $v_{i_r} \in B$. Define $\|x\| = \sum_{r=1}^n |\alpha_{i_r}|$. Then $\| \cdot \|$ is a norm on V. We will consider the triangle inequality. We will not give a complete proof of it, but we will mention by an example how we can proceed for a general proof.

Let $x, y \in V$. We can assume that both of x and y are nonzero. Suppose $x = \alpha_{i_1} v_{i_1} + \alpha_{i_2} v_{i_2} + \alpha_{i_3} v_{i_3} + \alpha_{i_4} v_{i_4}$ and $y = \beta_{j_1} v_{j_1} + \beta_{j_2} v_{j_2} + \beta_{j_3} v_{j_3}$. First, suppose that no v_{i_r} is equal to any of v_{j_s}. Then

$$x + y = \alpha_{i_1} v_{i_1} + \alpha_{i_2} v_{i_2} + \alpha_{i_3} v_{i_3} + \alpha_{i_4} v_{i_4} + \beta_{j_1} v_{j_1} + \beta_{j_2} v_{j_2} + \beta_{j_3} v_{j_3}.$$

Note that this is a unique representation of $x + y$ as a linear combination of elements of B. Then

$$\|x + y\| = \sum_{r=1}^4 |\alpha_{i_r}| + \sum_{s=1}^3 |\beta_{j_s}| = \|x\| + \|y\|.$$

Now suppose that $v_{i_3} = v_{j_2}$, $v_{i_4} = v_{j_3}$, and other are distinct. Then

$$x + y = (\alpha_{i_1} v_{i_1} + \alpha_{i_2} v_{i_2}) + \beta_{j_1} v_{j_1} + (\alpha_{i_3} + \beta_{j_2})v_{i_3} + (\alpha_{i_4} + \beta_{j_3})v_{i_4}.$$

This shows that

$$\begin{aligned}\|x + y\| &= |\alpha_{i_1}| + |\alpha_{i_2}| + |\beta_{j_1}| + |\alpha_{i_3} + \beta_{j_1}| + |\alpha_{i_4} + \beta_{j_1}| \\ &\leq |\alpha_{i_1}| + |\alpha_{i_2}| + |\beta_{j_1}| + |\alpha_{i_1}| + |\beta_{j_1}| + |\alpha_{i_4}| + |\beta_{j_1}| \\ &= \|x\| + \|y\|.\end{aligned}$$

The metric induced by a norm satisfies nice properties because of its algebraic structure compatible with norm. For example, we have the following:

Proposition 2.2.11. *Let V be a normed space, and let $x, y \in V$. If $B(x, r) \subseteq B(y, s)$, then $r \leq s$. Moreover, if $x \neq y$, then $r < s$.*

Proof. If $x = y$, then clearly $r \leq s$. Assume that $x \neq y$. Let $0 \leq t < r$. Consider $z = x + t\frac{(x-y)}{\|x-y\|}$. Then $\|z - x\| = t < r$. This shows that $z \in B(x, r)$. This implies that $z \in B(y, s)$. Therefore $\|z - y\| < s$. Putting the value of z, we get

$$\left\| (x - y) + t\frac{(x - y)}{\|x - y\|} \right\| < s.$$

In other words,

$$\|x - y\|\left(1 + \frac{t}{\|x - y\|} \right) = \|x - y\| + t < s.$$

Hence $t < s$. This shows that whenever $t < r$, we have $t < s$. Thus $r \leq s$.

Now assume that $x \neq y$. Then $\|x - y\| \neq 0$. On the contrary, assume that $r = s$. Since $x \in B(x, r)$, $\|x - y\| < r$. Choose $t \in \mathbb{R}$ such that

$$\frac{r}{\|x - y\|} - 1 < t < \frac{r}{\|x - y\|}.$$

Let $z = x - t(y - x)$. Then $\|z - x\| < r$ and $\|z - y\| > r$. This shows that $z \in B(x, r)$ but $z \notin B(y, r)$. This is a contradiction. □

Corollary 2.2.12. *Let V be a normed space, and let $x, y \in V$. Then $B(x, r) = B(y, s)$ if and only if $x = y$ and $r = s$.*

Proof. Suppose that $B(x, r) = B(y, s)$. If the statement does not hold, then either $x \neq y$ or $r \neq s$. If $x \neq y$, then by Proposition 2.2.11 we have $r < s$ and $s < r$, which is not possible. Now suppose that $r \neq s$. Further, assume that $r < s$. Let $z = x + t\frac{(x-y)}{\|x-y\|}$, where $r < t < s$. Then $z \in B(y, s)$, but $z \notin B(x, r)$. This is a contradiction. □

Let us see another property of the norm on a vector space V. A nonempty subset A of V is called convex if $\alpha x + (1 - \alpha)y \in A$ for all $x, y \in A$ and $0 \leq \alpha \leq 1$. It is easy to observe that a unit open ball $B(0, 1) = \{x \in V \mid \|x\| < 1\}$ is convex. Indeed, its converse is also true in the following sense.

Proposition 2.2.13. *Let $v : V \to \mathbb{R}$ be a map satisfying the following conditions:*

(i) $v(x) \geq 0$,
(ii) $v(x) = 0$ *if and only if* $x = 0$,
(iii) $v(\alpha x) = |\alpha|v(x)$.

Suppose that the set $B = \{x \in V \mid v(x) < 1\}$ is convex. Then the map v is a norm on V.

Proof. To show that v is a norm on V, we have to show that v satisfies the triangle inequality. If any of x or y is zero, then it is clearly satisfied. Suppose that $x \neq 0$ and $y \neq 0$. Let $u = \frac{x}{tv(x)}$ and $v = \frac{y}{tv(y)}$, where $t > 1$. Then $v(u) = v(v) < 1$. Since B is convex,

$$v\left(\frac{v(x)}{v(x) + v(y)}u + \frac{v(x)}{v(y) + v(y)}v\right) < 1.$$

This implies that $v(x+y) < t(v(x)+v(y))$. Since $t > 1$ is arbitrary, $v(x+y) \leq v(x)+v(y)$. $\qquad \square$

Let V_1 and V_2 be normed spaces over the field $\mathbb{F}$. Let $T : V_1 \to V_2$ be a linear transformation such that there exists a positive real number k such that $\|T(x)\| \leq k\|x\|$ for all $x \in V_1$. Let us collect all such linear transformations in a set $B(V_1, V_2)$. Define $\|\cdot\| : B(V_1, V_2) \to \mathbb{R}$ by

$$\|T\| = \inf\{k \mid \|T(x)\| \leq k\|x\| \text{ for all } x \in V_1\}.$$

We can check that $\|\cdot\|$ is a norm on $B(V_1, V_2)$. Note that $\|T(x)\| \leq \|T\|\|x\|$ for all $x \in V_1$.

Proposition 2.2.14. *Let $T \in B(V_1, V_2)$. Then*

$$\|T\| = \sup\{\|T(x)\| \mid \|x\| \leq 1\}.$$

Proof. If $\|x\| \leq 1$, then $\|T(x)\| \leq \|T\|\|x\| \leq \|T\|$. This implies that

$$\sup\{\|T(x)\| \mid \|x\| \leq 1\} \leq \|T\|.$$

Let $\epsilon > 0$. Then there is $y \neq 0$ such that

$$\|T(y)\| > (\|T\| - \epsilon)\|y\|.$$

Let $z = \frac{y}{\|y\|}$. Then we have

$$\|T(z)\| = \frac{\|T(y)\|}{\|y\|} > \|T\| - \epsilon.$$

This implies that

$$\sup\{\|T(x)\| \mid \|x\| \leq 1\} \geq \|T(z)\| > \|T\| - \epsilon.$$

Hence

$$\sup\{\|T(x)\| \mid \|x\| \leq 1\} \geq \|T\|. \qquad \square$$

Remark 2.2.15. The linear transformation $T : V_1 \to V_2$ satisfying the above condition is called a bounded linear transformation. We are avoiding this terminology here because this is different from a bounded map between the metric spaces X and Y. A map $f : X \to Y$ is called a bounded map if the image set Im f is a bounded set in Y. Observe that the identity map $I : \mathbb{R} \to \mathbb{R}$ satisfies the above condition but is not a bounded map.

Definition 2.2.16. Let V be vector space over a field $\mathbb{F}$. Then a map $\langle \cdot, \cdot \rangle : V \times V \to \mathbb{F}$ is called an inner product if for all $x, y, z \in V$ and $\alpha, \beta \in F$, the following conditions are satisfied:

(i) $\langle x, y \rangle \in \mathbb{R}$ and $\langle x, y \rangle \geq 0$,
(ii) $\langle x, x \rangle = 0$ if and only if $x = 0$,
(iii) $\langle \alpha x + \beta y, z \rangle = \alpha \langle x, z \rangle + \beta \langle y, z \rangle$,
(iv) $\langle x, y \rangle = \overline{\langle y, x \rangle}$.

If $\langle \cdot, \cdot \rangle$ is an inner product on a vector space V, then we say that the pair $(V, \langle \cdot, \cdot \rangle)$ is an inner product space. When an inner product is given on V, we will only say that V is an inner product space.

Example 2.2.17. Define $\langle \cdot, \cdot \rangle : \mathbb{F}^n \times \mathbb{F}^n \to \mathbb{F}$ by $\langle x, y \rangle = \sum_{k=1}^n x_k \overline{y_k}$, where $x = (x_1, \ldots, x_n)$ and $y = (y_1, \ldots, y_n)$. Then $\langle \cdot, \cdot \rangle$ is an inner product on $\mathbb{F}^n$. This inner product is called the standard or usual inner product on $\mathbb{F}^n$.

Let V be an inner product space. Define a map $\| \cdot \| : V \to \mathbb{R}$ by $\|x\| = +\sqrt{\langle x, x \rangle}$. We will observe that $\| \cdot \|$ is a norm on V. Clearly, $\| \cdot \|$ satisfies the first three conditions of the norm. To prove the triangle inequality, we need the following:

Proposition 2.2.18 (Cauchy–Schwarz inequality). *Let V be an inner product space, and let $\| \cdot \|$ be the map as defined above. Then, for all $x, y \in V$,*

$$|\langle x, y \rangle| \leq \|x\| \|y\|.$$

The equality holds if and only if one of x or y is a scalar multiple of the other.

Proof. If any of x or y is zero, then the above inequality is satisfied. Suppose that $x \neq 0$ and $y \neq 0$. Write

$$x = \frac{\langle x, y \rangle}{\|y\|^2} y + z,$$

where $z = x - \frac{\langle x, y \rangle}{\|y\|^2} y$. Note that

$$\left\langle \frac{\langle x, y \rangle}{\|y\|^2} y, z \right\rangle = 0.$$

Now

$$\|x\|^2 = \langle x, x \rangle$$

$$= \left\langle \frac{\langle x, y \rangle}{\|y\|^2} y + z, \frac{\langle x, y \rangle}{\|y\|^2} y + z \right\rangle$$

$$= \left| \frac{\langle x, y \rangle}{\|y\|^2} y \right|^2 + \|z\|^2$$

$$= \frac{|\langle x, y \rangle|^2}{\|y\|^2} + \|z\|^2$$

$$\geq \frac{|\langle x, y \rangle|^2}{\|y\|^2}.$$

Therefore

$$\left| \langle x, y \rangle \right| \leq \|x\| \|y\|.$$

Note that the equality holds if and only if $z = 0$. This equivalently means that

$$x = \frac{\langle x, y \rangle}{\|y\|^2} y. \qquad \square$$

Now observe that

$$\|x + y\|^2 = \langle x + y, x + y \rangle = \|x\|^2 + 2\operatorname{Re}\langle x, y \rangle + \|y\|^2$$

$$\leq \|x\|^2 + 2|\langle x, y \rangle| + \|y\|^2$$

$$\leq \|x\|^2 + 2\|x\|\|y\| + \|y\|^2$$

$$= (\|x\| + \|y\|)^2.$$

Therefore $\|x + y\| \leq \|x\| + \|y\|$. Thus an inner product induces a norm on V, but a norm may be not induced by an inner product. Observe that if V is an inner product space and $\| \cdot \|$ is a norm induced by the inner product on V, then for all $x, y \in V$, we have

$$\|x + y\|^2 + \|x - y\|^2 = (\|x\|^2 + \|y\|^2).$$

This identity is called the parallelogram identity.

Take $x = (1, 0, \ldots, 0, \ldots)$ and $y = (0, 1, \ldots, 0, \ldots)$ in the space ℓ_p. We can see that the parallelogram identity is not satisfied if $p \neq 2$. This shows that the norm on the space ℓ_p ($p \neq 2$) is not induced by an inner product.

2.3 Metric-preserving maps

Let us revisit Examples 2.1.8 and 2.1.9. Let d be a metric on a set X. Then by condition (i) of Definition 2.1.1, d is a map from $X \times X$ to $[0, \infty)$. Consider the map $f : [0, \infty) \to [0, \infty)$

defined by $f(x) = \frac{x}{1+x}$. By Example 2.1.8 the composition $f \circ d$ is a metric on X. We can similarly deal with Example 2.1.9.

Definition 2.3.1. A map $f : [0, \infty) \to [0, \infty)$ is called a metric-preserving map if $f \circ d$ is a metric for all metric spaces (X, d).

Example 2.3.2. The map $f : [0, \infty) \to [0, \infty)$ defined by $f(x) = x^2$ is not a metric-preserving map. Consider the usual metric d on $\mathbb{R}$. Then

$$(f \circ d)(1,3) > (f \circ d)(1,2) + (f \circ d)(2,3).$$

Example 2.3.3. Define the map $f : [0, \infty) \to [0, \infty)$ by

$$f(x) = \begin{cases} 0 & \text{if } x = 0, \\ 1 & \text{if } x > 0. \end{cases}$$

Then $f \circ d$ is the discrete metric.

Definition 2.3.4. A map $f : [0, \infty) \to [0, \infty)$ is called subadditive if $f(x+y) \le f(x) + f(y)$ for all $x, y \in [0, \infty)$.

Proposition 2.3.5. *Let $f : [0, \infty) \to [0, \infty)$ be a metric-preserving map. Then f is subadditive.*

Proof. Let d be the usual metric on $\mathbb{R}$. Let $x, y, z \in [0, \infty)$. Since f is a metric-preserving map,

$$\begin{aligned} f(x+y) &= (f \circ d)(0, x+y) \\ &\le (f \circ d)(0, x) + f \circ d(x, x+y) \\ &= f(x) + f(y). \end{aligned}$$
$\square$

Definition 2.3.6. A map $f : [0, \infty) \to [0, \infty)$ is called amenable if $f^{-1}\{0\} = \{0\}$.

Proposition 2.3.7. *Let $f : [0, \infty) \to [0, \infty)$ be a metric-preserving map. Then f is amenable.*

Proof. Note that if $f(0) \ne 0$, then $f \circ d$ is not a metric. Also, suppose there is $a > 0$ such that $f(a) = 0$. Take the usual metric d on $\mathbb{R}$. Then $(f \circ d)(0, a) = f(a) = 0$, but $a \ne 0$. This shows that $f \circ d$ is not a metric. $\square$

It may happen that $f : [0, \infty) \to [0, \infty)$ is a subadditive and amenable but not a metric-preserving map. For example, we will see later that $f(x) = \frac{x}{1+x^2}$ is a subadditive and amenable but not a metric-preserving map. We have a sufficient condition for f to be a metric-preserving map.

Proposition 2.3.8. *Let $f : [0, \infty) \to [0, \infty)$ be a subadditive, amenable, and increasing map. Then f is a metric-preserving map.*

Proof. Let (X, d) be a metric space, and let $x, y, z \in X$. We will only prove the triangle inequality for $f \circ d$. Note that

$$d(x, y) \le d(x, z) + d(z, y).$$

Since f is increasing,

$$f(d(x, y)) \le f(d(x, z) + d(z, y)).$$

Since f is subbaditive,

$$f(d(x, z) + d(z, y)) \le f(d(x, z)) + f(d(z, y)).$$

This shows that $f \circ d$ satisfies the triangle inequality. ☐

Definition 2.3.9. Let $a, b, c \in [0, \infty)$. Then the triple (a, b, c) is called a triangle triple if $a \le b + c$, $b \le a + c$, and $c \le a + b$.

Observe that (a, b, c) is a triangle triple if and only if $|a - b| \le c \le a + b$.

Proposition 2.3.10 (Borsik and Dobos). *Let $f : [0, \infty) \to [0, \infty)$ be an amenable map. Then f is a metric-preserving map if and only if $(f(a), f(b), f(c))$ is a triangle triple for all triangle triples (a, b, c).*

Proof. Let (a, b, c) be a triangle triple. Let d be the usual metric on $\mathbb{R}^2$. Then there are $x, y, z \in \mathbb{R}$ such that $d(x, y) = a$, $d(y, z) = b$, and $d(x, z) = c$. Since f is metric preserving, $(f(a), f(b), f(c))$ is a triangle triple.

Conversely, suppose that $(f(a), f(b), f(c))$ is a triangle triple for all triangle triples (a, b, c). Let (X, d) be a metric space. Since $(d(x, y), d(y, z), d(x, z))$ is a triangle triple, $f \circ d$ satisfies the triangle inequality. We can easily check that $f \circ d$ satisfies the other conditions of a metric. ☐

Note 2.3.11. Let (a, b, c) be a triangle triple with $a > 0$, and let d be the usual metric on $\mathbb{R}^2$. Let $x = (\frac{a}{2}, 0)$, $y = (-\frac{a}{2}, 0)$, and $z = (u, v)$, where $u = \frac{b^2 - c^2}{2a}$, and v is given by $x^2 + \frac{a^2}{4} + ax + y^2 = b^2$. Check that for $x, y, z \in \mathbb{R}^2$, we have $d(x, y) = a$, $d(y, z) = b$, and $d(x, z) = c$.

Proposition 2.3.12 (Sreenivasan and Terpe). *Let $f : [0, \infty) \to [0, \infty)$ be an amenable map. Then f is a metric-preserving map if and only if $f(a) \le f(b) + f(c)$ for all triangle triples (a, b, c).*

Proof. Suppose that f is a metric preserving map. Let (a, b, c) be a triangle triple. Let d be the usual metric on $\mathbb{R}^2$. Then there are $x, y, z \in \mathbb{R}^2$ such that $d(x, y) = a$, $d(y, z) = b$, and $d(x, z) = c$. Since f is metric preserving,

$$f(a) = f(d(x,y)) \leq f(d(x,z)) + f(d(z,y)) = f(b) + f(c).$$

By similar arguments as in the proof of Proposition 2.3.10, the converse follows. □

Proposition 2.3.13. *Let $f : [0,\infty) \to [0,\infty)$ be a metric-preserving map. Then the following statements are equivalent:*

(i) *f is continuous,*

(ii) *f is continuous at 0,*

(iii) *for each $\epsilon > 0$, there is $x > 0$ such that $f(x) < \epsilon$.*

Proof. We can easily observe that (i) $\Rightarrow$ (ii) and (ii) $\Rightarrow$ (iii). We will prove (iii) $\Rightarrow$ (i).

Suppose (iii) holds. Let $a \in [0,\infty)$ and $\epsilon > 0$. By (iii) there is $h > 0$ such that $f(h) < \epsilon$. Note that $(h, a, a+h)$ is a triangle triple. Therefore $(f(h), f(a), f(a+h))$ is a triangle triple. This implies that

$$|f(a+h) - f(a)| \leq f(h) < \epsilon.$$

This shows that f is continuous at a. Since $a \in [0,\infty)$ is arbitrary, f is continuous. □

Proposition 2.3.14. *Let $f : [0,\infty) \to [0,\infty)$ be a metric-preserving map. Then, for each $\delta > 0$, there is $\epsilon > 0$ such that*

$$x \geq \delta \Rightarrow f(x) \geq \epsilon.$$

Proof. On the contrary, assume that there is $\delta > 0$ such that for each $\epsilon > 0$,

$$x \geq \delta \Rightarrow f(x) < \epsilon.$$

Choose $a \geq \delta > 0$ such that $f(a) < \frac{f(\delta)}{2}$. Note that (a, a, δ) is a triangle triple but $(f(a), f(a), f(\delta))$ is not. This is a contradiction. □

Corollary 2.3.15. *Let $f : [0,\infty) \to [0,\infty)$ be a metric-preserving map. Then $\lim_{x \to \infty} f(x) \neq 0$.*

Observe that $\lim_{x \to \infty} \frac{x}{1+x^2} = 0$. This shows that $f(x) = \frac{x}{1+x^2}$ is not a metric-preserving map.

Proposition 2.3.16. *Let $f : [0,\infty) \to [0,\infty)$ be amenable such that for all $x > 0$, $v \leq f(x) \leq 2v$ for some $v > 0$. Then f is a metric-preserving map.*

Proof. Let $v > 0$ be such that $v \leq f(x) \leq 2v$ for all $x > 0$. Let (a, b, c) be a triangle triple. We will show that $f(a) \leq f(b) + f(c)$. If any one of a, b, or c is zero, then it is clearly satisfied. Suppose that $a \neq 0$, $b \neq 0$, and $c \neq 0$. Then

$$f(a) \leq 2v = v + v \leq f(b) + f(c).$$

By Proposition 2.3.12, f is a metric-preserving map. □

Example 2.3.17. Define the map $f : [0, \infty) \to [0, \infty)$ by

$$f(x) = \begin{cases} 0 & \text{if } x = 0, \\ 1 & \text{if } x \in (0, \infty) \cap \mathbb{Q}, \\ 2 & \text{if } x \in (0, \infty) \cap (\mathbb{R} \setminus \mathbb{Q}). \end{cases}$$

By Proposition 2.3.16, f is a metric-preserving map. Note that f is discontinuous at each point of $[0, \infty)$. Also, note that f is not increasing.

2.4 Open and closed sets

Let us recall the definition of continuity of a map $f : \mathbb{R} \to \mathbb{R}$ from elementary calculus. We note that the continuity depends on the open intervals. We would like to generalize the notion of an open interval to the so-called open sets and try to get a notion of continuity in the next chapter, which does not depend on the distance.

Definition 2.4.1. Let A be a set in a metric space X. Then a point $x \in X$ is called an interior point of A if there is a positive real number r such that $B(x, r) \subseteq A$.

We denote the set of all interior points of A by $A°$ or $\text{Int}_X A$. When the metric space X is given, we will denote it by $\text{Int} A$. Note that if A is nonempty and $x \in X$ is an interior point of A, then $x \in A$.

Definition 2.4.2. Let A be a set in a metric space X. Then a point $x \in X$ is called an exterior point of A if there is a positive real number r such that $B(x, r) \subseteq X \setminus A$.

We denote the set of all exterior points of A by $\text{Ext}_X A$ or $\text{Ext} A$.

Definition 2.4.3. Let A be a set in a metric space X. Then a point $x \in X$ is called a boundary point of A if it is neither an interior point nor an exterior point of A.

We denote the set of all boundary points of A by $\text{Bd}_X A$ or $\text{Bd} A$. Note that if $x \in X$ is a boundary point of A, then for all $r > 0$,

$$B(x, r) \cap A \neq \emptyset \quad \text{and} \quad B(x, r) \cap (X \setminus A) \neq \emptyset.$$

Note that for any set A in a metric space X, $\text{Int} A$, $\text{Ext} A$, and $\text{Bd} A$ forms a partition of X. Also, note that $\text{Int} A = \text{Ext}(X \setminus A)$, $\text{Ext} A = \text{Int}(X \setminus A)$, and $\text{Bd} A = \text{Bd}(X \setminus A)$.

Example 2.4.4. Consider a closed interval $[a, b]$ in the real line. Let $x \in (a, b)$. Let $r = \min\{|x - a|, |x - b|\}$. Then $B(x, r) = (x - r, x + r) \subseteq [a, b]$. Also, note that no point of $\mathbb{R} \setminus (a, b)$ can be an interior point of $[a, b]$. This shows that $\text{Int}[a, b] = (a, b)$. We can similarly observe that $\text{Ext}[a, b] = (-\infty, a) \cup (b, \infty)$ and $\text{Bd}[a, b] = \{a, b\}$.

Example 2.4.5. Consider the set $\mathbb{Q}$ in the real line. Note that any open interval around a point contains irrational numbers. This implies that Int $\mathbb{Q} = \emptyset$. We can similarly show that Int $\mathbb{R} \setminus \mathbb{Q} =$ Ext $\mathbb{Q} =$ Ext $\mathbb{R} \setminus \mathbb{Q} = \emptyset$. Also, observe that Bd $\mathbb{Q} =$ Bd $\mathbb{R} \setminus \mathbb{Q} = \mathbb{R}$.

Definition 2.4.6. A set in a metric space X is called an open set if every point of A is an interior point of A.

Consider the Euclidean space $\mathbb{R}^n$. Let $a_1, \ldots, a_n$ be arbitrary but fixed positive real numbers. We can check that

$$\{(x_1, \ldots, x_n) \in \mathbb{R}^n \mid |x_i| < a_i\}$$

is an open set in $\mathbb{R}^n$. The same situation may not be true in an infinite-dimensional normed space.

Proposition 2.4.7. *Let (a_n) be an arbitrary but fixed sequence of positive real numbers. Then the set*

$$U = \{(x_n) \in \ell_2 \mid |x_i| < a_i \text{ for all } i\}$$

is an open set in ℓ_2 if and only if

$$\inf\{a_n \mid n \in \mathbb{N}\} > 0.$$

Proof. Suppose that U is an open set in ℓ_2. Since $a_1 > 0$, there is $\epsilon > 0$ such that $a_1 > \epsilon > 0$. Then

$$x = (x_n) = (a_1 - \epsilon, 0, 0, \ldots) \in U.$$

Since U is an open set, there is $\delta > 0$ such that

$$B(x, \delta) \subseteq U.$$

Let $r = \min\{\epsilon, \frac{\delta}{2}\}$. Then

$$D(x, r) \subseteq B(x, \delta) \subseteq U.$$

Let $m \geq 2$. Consider the sequence $y = (y_n)$ defined by

$$y_n = \begin{cases} a_1 - \epsilon & \text{if } n = 1, \\ r & \text{if } n = m, \\ 0 & \text{otherwise.} \end{cases}$$

Note that $y \in D(x, r) \subseteq U$. This implies that $r < a_m$. Observe that $a_1 > \epsilon \geq r$. Since $m \geq 2$ is arbitrary, $a_n \geq r$ for all $n \in \mathbb{N}$. This shows that $\inf\{a_n \mid n \in \mathbb{N}\} \geq r > 0$.

Conversely, suppose that $\inf\{a_n \mid n \in \mathbb{N}\} > 0$. Then there is $s > 0$ such that $a_n > s$ for all $n \in \mathbb{N}$. Let $x = (x_n) \in U$. We will show that x is an interior point of U.

Since $\sum_{n=1}^{\infty} |x_n|^2 < \infty$, there is a positive integer k such that

$$\sum_{n=k+1}^{\infty} |x_n|^2 < \frac{s^2}{4}.$$

Let $r = \min\{\frac{s}{2}, a_1 - |x_1|, \ldots, a_k - |x_k|\}$. Since $x \in U$, $r > 0$. Let $y = (y_n) \in B(x,r)$. Then $\|y - x\| < r$. For $1 \le n \le k$, we have

$$\begin{aligned}
|y_n| &\le |y_n - x_n| + |x_n| \\
&\le \|y - x\| + |x_n| \\
&< r + |x_n| \\
&\le a_n - |x_n| + |x_n| \\
&= a_n.
\end{aligned}$$

Let $x' = (x_{n+k})$ and $y' = (y_{n+k})$. Then

$$\begin{aligned}
\|y'\| &\le \|y' - x'\| + \|x'\| \\
&\le \|y - x\| + \left(\sum_{n=k+1}^{\infty} |x_n|^2\right)^{\frac{1}{2}} \\
&< r + \frac{s}{2} \\
&\le s.
\end{aligned}$$

This implies that for $n \ge k + 1$, we have

$$\begin{aligned}
|y_n| &\le \|y'\| \\
&< s \\
&< a_n.
\end{aligned}$$

Therefore $|y_n| < a_n$ for all $n \in \mathbb{N}$. This shows that $y = (y_n) \in U$. Since $y \in B(x,r)$ is arbitrary, $B(x,r) \subseteq U$. Hence x is an interior point of U. Since $x \in U$ is arbitrary, U is open. $\square$

Let A be a set in a metric space X. Note that $\operatorname{Int} A$ is the largest open set contained in A. In other words,

$$\operatorname{Int} A = \bigcup\{U_\alpha \mid U_\alpha \text{ is open in } X \text{ contained in } A\}.$$

Also, A is open in X if and only if $\operatorname{Int} A = A$. Observe that if $A \subseteq B$, then $\operatorname{Int} A \subseteq \operatorname{Int} B$. For this, note that $\operatorname{Int} A \subseteq A \subseteq B$. Since $\operatorname{Int} B$ is the largest open set contained in B, $\operatorname{Int} A \subseteq \operatorname{Int} B$.

Proposition 2.4.8. *Let $\{A_\alpha \mid \alpha \in \mathcal{I}\}$ be a family of sets in a metric space. Then*
(i) $\operatorname{Int}(\bigcap_{\alpha \in \mathcal{I}} A_\alpha) \subseteq \bigcap_{\alpha \in \mathcal{I}} \operatorname{Int} A_\alpha;$
(ii) *if $\mathcal{I}$ is finite, then* $\operatorname{Int}(\bigcap_{\alpha \in \mathcal{I}} A_\alpha) = \bigcap_{\alpha \in \mathcal{I}} \operatorname{Int} A_\alpha;$
(iii) $\bigcup_{\alpha \in \mathcal{I}} \operatorname{Int} A_\alpha \subseteq \operatorname{Int}(\bigcup_{\alpha \in \mathcal{I}} A_\alpha).$

Proof. (i) Note that $\bigcap_{\alpha \in \mathcal{I}} A_\alpha \subseteq A_\alpha$ for all $\alpha \in \mathcal{I}$. Then

$$\operatorname{Int}\left(\bigcap_{\alpha \in \mathcal{I}} A_\alpha\right) \subseteq \operatorname{Int} A_\alpha \quad \text{for all } \alpha \in \mathcal{I}.$$

This implies that

$$\operatorname{Int}\left(\bigcap_{\alpha \in \mathcal{I}} A_\alpha\right) \subseteq \bigcap_{\alpha \in \mathcal{I}} \operatorname{Int} A_\alpha.$$

(ii) Without loss of generality, assume that $\mathcal{I} = \{1, \ldots, n\}$. Let $U = \bigcap_{\alpha=1}^{n} \operatorname{Int} A_\alpha$. Note that U is open. Since $\operatorname{Int} A_\alpha \subseteq A_\alpha$,

$$U = \bigcap_{\alpha=1}^{n} \operatorname{Int} A_\alpha \subseteq \bigcap_{\alpha=1}^{n} A_\alpha.$$

Since for any set B, $\operatorname{Int} B$ is the largest open set contained in B,

$$\bigcap_{\alpha=1}^{n} \operatorname{Int} A_\alpha \subseteq \operatorname{Int}\left(\bigcap_{\alpha=1}^{n} A_\alpha\right).$$

(iii) Left as an exercise. □

Remark 2.4.9.
(i) If $\mathcal{I}$ is not a finite set, then the equality in Proposition 2.4.8(i) need not hold. For example, for all $n \in \mathbb{N}$, consider the sets $A_n = (-\frac{1}{n}, \frac{1}{n})$ in the real line. Then

$$\bigcap_{n \in \mathbb{N}} \operatorname{Int} A_n = \{0\} \quad \text{and} \quad \operatorname{Int}\left(\bigcap_{n \in \mathbb{N}} A_n\right) = \emptyset.$$

(ii) Consider the singletons $\{a\}$ in the real line. Note that $\operatorname{Int}\{a\} = \emptyset$ and $\bigcup_{a \in \mathbb{R}} \{a\} = \mathbb{R}$. This implies that

$$\emptyset = \bigcup_{a \in \mathbb{R}} \operatorname{Int}\{a\} \subseteq \operatorname{Int} \bigcup_{a \in \mathbb{R}} \{a\} = \mathbb{R}.$$

This shows that the equality in Proposition 2.4.8(iii) need not hold. This is even not true for a finite union. For example,

$$\emptyset = \operatorname{Int} \mathbb{Q} \cup \operatorname{Int} \mathbb{R} \setminus \mathbb{Q} \subseteq \operatorname{Int}(\mathbb{Q} \cup \mathbb{R} \setminus \mathbb{Q}) = \mathbb{R}.$$

Proposition 2.4.10. *An open ball in a metric space is an open set.*

Proof. Consider an open ball $B(x,r)$ in a metric space X. Let $y \in B(x,r)$. If $y = x$, then $B(y,r) \subseteq B(x,r)$. Now suppose that $y \neq x$. Then $s = d(x,y) > 0$. Let $\epsilon = \min\{s, r-s\}$. We claim that $B(y,\epsilon) \subseteq B(x,r)$. For this, let $z \in B(y,\epsilon)$. This implies that $d(y,z) < \epsilon$. Then

$$d(z,x) \leq d(z,y) + d(y,x)$$
$$< \epsilon + s$$
$$\leq r - s + s$$
$$= r.$$

Hence $z \in B(x,r)$. □

Definition 2.4.11. A set A in a metric space X is called a closed set if $X \setminus A$ is open in X.

Example 2.4.12. Each singleton set in a metric space X is a closed set. For this, let $x \in X$. Let $y \in X$ be such that $y \neq x$. Then $r = d(x,y) > 0$. Note that $x \notin B(y,r)$. This equivalently means that $B(y,r) \subseteq X \setminus \{x\}$. Hence $X \setminus \{x\}$ is open in X.

Proposition 2.4.13. *A closed ball in a metric space is a closed set.*

Proof. Consider a closed ball $D(x,r)$ in a metric space X. We will prove that $X \setminus D(x,r)$ is open in X. If $D(X,r) = X$, then $X \setminus D(x,r) = \emptyset$. Since $\emptyset$ contains no point, $\emptyset$ is open. Now assume that $D(X,r) \neq X$. Let $y \in X \setminus D(x,r)$. Then $s = d(x,y) > 0$. We claim that $B(y, s-r) \subseteq X \setminus D(x,r)$. For this, let $z \in B(y, s-r)$. This implies that $d(y,z) < s-r$. Then

$$s = d(x,y) \leq d(x,z) + d(z,y)$$
$$< d(x,z) + s - r.$$

Therefore $d(x,z) > r$. Hence $z \in X \setminus D(x,r)$. □

The interior of a closed ball in a metric space need not be an open ball. For this, consider the following example.

Example 2.4.14. Let X be the discrete metric space containing at least two elements. Since $D(x,1) = X$ for each $x \in X$, $\operatorname{Int} D(x,1) = X$. Note that $B(x,1) = \{x\}$.

Proposition 2.4.15. *Let V be a normed space, and let $x \in V$. Then $\operatorname{Int} D(x,r) = B(x,r)$ for all $r > 0$.*

Proof. Since $B(x,r) \subseteq D(x,r)$, $B(x,r) \subseteq \operatorname{Int} D(x,r)$. For the converse, let $y \notin B(x,r)$. Then $\|x-y\| \geq r$. If $\|x-y\| > r$, then $y \notin D(x,r)$. This implies that $y \notin \operatorname{Int} D(x,r)$. Let $\|x-y\| = r$. For each $\epsilon > 0$, consider an open ball $B(y,\epsilon)$ around y. Let

$$z = y + \frac{\epsilon}{2} \frac{y - x}{\|x - y\|}.$$

Then $\|z - y\| = \frac{\epsilon}{2} < \epsilon$, but $\|x - z\| = r + \frac{\epsilon}{2} > r$. This shows that $y \notin \operatorname{Int} D(x, r)$. □

Now we prove an important property of open sets in a metric space, which will prompt us to define a topology.

Theorem 2.4.16. *Let X be a metric space. Then*
(i) $\emptyset$ *and X are open sets;*
(ii) *if $\{U_\alpha \mid \alpha \in \mathcal{I}\}$ is a family of open sets of X, then $\bigcup_{\alpha \in \mathcal{I}} U_\alpha$ is open in X;*
(iii) *if $U_1, U_2, \ldots, U_n$ are open sets of X, then $U_1 \cap U_2 \cap \cdots \cap U_n$ is open in X.*

Proof. (i) We have already seen that $\emptyset$ is open in X. Since every open ball $B(x, r)$ around each $x \in X$ is contained in X, X is open in X.

(ii) Let $x \in \bigcup_{\alpha \in \mathcal{I}} U_\alpha$. Then, $x \in U_\beta$ for some $\beta \in \mathcal{I}$. Since U_β is open, there is $r > 0$ such that

$$B(x, r) \subseteq U_\beta \subseteq \bigcup_{\alpha \in \mathcal{I}} U_\alpha.$$

This shows that $\bigcup_{\alpha \in \mathcal{I}} U_\alpha$ is open in X.

(iii) It is sufficient to prove that $U_1 \cap U_2$ is open in X. Let $x \in U_1 \cap U_2$. Since U_1 and U_2 are open, there are $r_i > 0$, $i = 1, 2$, such that $B(x, r_i) \subseteq U_i$. Let $r = \min\{r_1, r_2\}$. Then $B(x, r) \subseteq U_1 \cap U_2$. □

Example 2.4.17. Let X be the discrete metric space. Since each singletons are open balls for suitable radius, the singletons are open sets in the discrete metric space. This shows that each set in the discrete metric space is open. This implies that each set in the discrete metric space is a closed set.

Example 2.4.18. Consider the set $\mathbb{Z}$ of integers in the real line. Since

$$\mathbb{R} \setminus \mathbb{Z} = \bigcup_{n \in \mathbb{Z}} (n, n + 1),$$

$\mathbb{Z}$ is closed in $\mathbb{R}$.

Example 2.4.19. Let $a \in \mathbb{R}$. Since $(a, \infty) = \bigcup_{n \in \mathbb{N}} (a, n)$, (a, ∞) is open in $\mathbb{R}$. Similarly, $(-\infty, a)$ is open in $\mathbb{R}$.

As an application De Morgan's law in Theorem 2.4.16, we get the following:

Theorem 2.4.20. *Let X be a metric space. Then*
(i) $\emptyset$ *and X are closed sets;*
(ii) *if $\{F_\alpha \mid \alpha \in \mathcal{I}\}$ is a family of closed sets of X, then $\bigcap_{\alpha \in \mathcal{I}} F_\alpha$ is closed in X;*
(iii) *if $F_1, F_2, \ldots, F_n$ are open sets of X, then $F_1 \cup F_2 \cup \cdots \cup F_n$ is closed in X.*

Example 2.4.21. Consider the real line. Note that $\bigcap_{n\in\mathbb{N}}(-\frac{1}{n},\frac{1}{n}) = \{0\}$ and singletons are not open in $\mathbb{R}$. This shows that the arbitrary intersection of open sets need not be open. By De Morgan's law we have $\bigcup_{n\in\mathbb{N}}\mathbb{R}\backslash(-\frac{1}{n},\frac{1}{n}) = \mathbb{R}\backslash\{0\}$. This also shows that the arbitrary union of closed sets need not be closed.

Definition 2.4.22. A set A in a metric space X is called a clopen set if it is both open and closed in X.

Example 2.4.23. The empty set $\emptyset$ and X are clopen sets in a metric space X.

Example 2.4.24. In the discrete space X, all sets are clopen.

Now we describe the open sets in terms of open balls in a metric space.

Proposition 2.4.25. *Let X be a metric space. Then a set U in X is open if and only if U is a union of open balls in X.*

Proof. If U is a union of open balls, then U is an open set. Conversely, suppose that U is open in X. If $U = \emptyset$, then trivially it is a union of sets over an empty indexing set. If $U \neq \emptyset$, then for each $x \in U$, there is an open ball $B(x, r_x)$ contained in U. This shows that $U = \bigcup_{x\in U} B(x, r_x)$. $\qquad\square$

We now describe the open sets in the real line.

Theorem 2.4.26. *A nonempty open set in the real line is the countable union of disjoint open sets.*

Proof. Let U be a nonempty open set in the real line. Let $V = U \cap \mathbb{Q}$. Note that V is nonempty and countable. For each $x \in V$, let I_x denote the union of all open intervals that are contained in U and contain x. Then I_x is the largest open interval contained in U that contains x.

Let $x, y \in V$. We claim that $I_x = I_y$ or $I_x \cap I_y = \emptyset$. For this, suppose that $I_x \neq I_y$ and $I_x \cap I_y \neq \emptyset$. Then $I_x \cup I_y$ is an open interval containing x and y. Since I_x is the largest open interval containing x, this is a contradiction. Note that U is the union of open intervals and each open interval contains rational numbers. This shows that $U = \bigcup_{x\in V} I_x$. $\qquad\square$

Recall from Example 2.1.14 the subspace (A, d_A) of a metric space (X, d). Let $x \in A$. Let $B(x, r)$ and $B_A(x, r)$ denote the open balls around x in X and in A, respectively. Then

$$B(x, r) \cap A = \{y \in A \mid d(x, y) < r\}$$
$$= \{y \in A \mid d_A(x, y) < r\}$$
$$= B_A(x, r).$$

Now we describe the open and closed sets in a subspace.

Proposition 2.4.27. *Let A be a subspace of a metric space X. Then the open sets of A are the intersections of open sets X with A.*

Proof. Let V be an open set in A. If $V = \emptyset$, then $V = \emptyset \cap A$. Now suppose that $V \neq \emptyset$. Let $x \in V$. Then there is an open ball $B_A(x, r_x)$ in A that is contained in A. By the above observation,

$$B_A(x, r_x) = B(x, r_x) \cap A.$$

Now

$$V = \bigcup_{x \in A} B_A(x, r_x)$$
$$= \bigcup_{x \in A} (B(x, r_x) \cap A)$$
$$= \left(\bigcup_{x \in A} B(x, r_x) \right) \cap A.$$

Let $U = \bigcup_{x \in A} B(x, r_x)$. Then $V = U \cap A$. Observe that U is open in X as $B(x, r_x)$ is open in X.

Now let $V = U \cap A$, where U is open in X. If $V = \emptyset$, then it is open in A. Let $V \neq \emptyset$ and $x \in V$. Then $x \in U$. Since U is open in X, there is an open ball $B(x, r)$ around x contained in U. Hence

$$B(x, r) \cap A \subseteq U \cap A = V.$$

Thus V is open in A. $\qquad\square$

Example 2.4.28. Consider the set $\mathbb{N}$ of natural numbers in the real line. Note that

$$\{x\} = \left(x - \frac{1}{3}, x + \frac{1}{3} \right) \cap \mathbb{N}.$$

Therefore each singleton is open in the subspace $\mathbb{N}$ of $\mathbb{R}$. This implies that each set in $\mathbb{N}$ is open in $\mathbb{N}$.

Example 2.4.29. Consider the real line. Note that $[0, \frac{1}{2})$ is open in $[0, 1]$ since $[0, \frac{1}{2}) = (-\frac{1}{2}, \frac{1}{2}) \cap [0, 1]$. Also, note that $[0, \frac{1}{2})$ is not open in $\mathbb{R}$.

Proposition 2.4.30. *Let Y be a subspace of a metric space X, and let $A \subseteq Y$. Then $\mathrm{Int}_Y A = \mathrm{Int}_X(A \cup (X \setminus Y)) \cap Y$.*

Proof. Note that

$$\mathrm{Int}_Y A = \bigcup_\alpha \{V_\alpha \mid V_\alpha \text{ is open in } Y, \text{ and } V_\alpha \subseteq A\}$$
$$= \bigcup_\alpha \{U_\alpha \cap Y \mid U_\alpha \text{ is open in } X, \text{ and } U_\alpha \cap Y \subseteq A\}$$
$$= \bigcup_\alpha \{U_\alpha \mid U_\alpha \text{ is open in } X, \text{ and } U_\alpha \subseteq A \cup (X \setminus Y)\} \cap Y$$
$$= \mathrm{Int}_X(A \cup (X \setminus Y)) \cap Y. \qquad\square$$

Proposition 2.4.31. *Let A be a subspace of a metric space X. Then the closed sets of A are the intersections of closed sets X with A.*

Proof. Let F be closed set in A. Then $A \setminus F$ is open in A. By Proposition 2.4.27 there is an open set U in X such that $A \setminus F = U \cap A$. Note that $F = A \cap (X \setminus U)$. Then F is an intersection of a closed set in X with A.

Now let $F = A \cap C$, where C is closed in X. Since $X \setminus C$ is open in X, $A \setminus F = (X \setminus C) \cap A$ is open in A. Hence F is closed in A. □

Example 2.4.32. Since

$$(-\sqrt{2}, \sqrt{2}) \cap \mathbb{Q} = [-\sqrt{2}, \sqrt{2}] \cap \mathbb{Q},$$

$(-\sqrt{2}, \sqrt{2}) \cap \mathbb{Q}$ is open as well as closed in the subspace $\mathbb{Q}$ of the real line.

Proposition 2.4.33. *Let A be a closed set of a metric space X. Then a set B of A is closed in A if and only if B is closed in X.*

Proof. Let B be closed in A. Then $B = F \cap A$ for some closed set F of X. Since A is closed in X, B is closed in X. Conversely, let B be closed in X. Since $B \subseteq A$, $B = B \cap A$. Therefore B is closed in A. □

We can similarly prove the following:

Proposition 2.4.34. *Let A be a open set of a metric space X. Then a set B of A is open in A if and only if B is open in X.*

Consider a set A in a metric space X. Then X itself is a closed set containing A. Consider the collection

$$\{F_\alpha \mid F_\alpha \text{ is a closed set containing } A\}.$$

Now consider the intersection all such F_α. By Theorem 2.4.20 the intersection is a closed set. Indeed, this is the smallest closed set containing A.

Definition 2.4.35. Let A be a set in a metric space X. The smallest closed set in X containing A is called the closure of A.

We denote the closure of A by $\overline{A}$ or $\mathrm{Cl}_X A$. When the metric space X is given, we will denote it by $\mathrm{Cl} A$. Note that

$$\mathrm{Cl} A = \bigcap \{F_\alpha \mid F_\alpha \text{ is closed in } X \text{ containing } A\}.$$

Example 2.4.36. The smallest closed set containing the open interval (a, b) in the real line is $[a, b]$. Therefore $\mathrm{Cl}(a, b) = [a, b]$. Similarly, $\mathrm{Cl}[a, b) = \mathrm{Cl}(a, b] = \mathrm{Cl}[a, b] = [a, b]$.

Example 2.4.37. Consider the set $\mathbb{Q}$ of rational numbers in the real line. Let F be a closed set in $\mathbb{R}$ such that $\mathbb{Q} \subseteq F \subseteq \mathbb{R}$. Note that $\mathbb{Q}$ is not a closed set in $\mathbb{R}$. This implies that $F \neq \mathbb{Q}$.

Also, note that $\mathbb{R} \setminus F$ is open in $\mathbb{R}$ and $\mathbb{R} \setminus F \subseteq \mathbb{R} \setminus \mathbb{Q}$. Since $\operatorname{Int} \mathbb{R} \setminus \mathbb{Q} = \emptyset$, $\mathbb{R} \setminus F = \emptyset$. This shows that $F = \mathbb{R}$. Hence $\operatorname{Cl} \mathbb{Q} = \mathbb{R}$. We can similarly observe that $\operatorname{Cl}(\mathbb{R} \setminus \mathbb{Q}) = \mathbb{R}$.

Definition 2.4.38. A set A in a metric space X is called a dense set if $\operatorname{Cl} A = X$.

By Example 2.4.37, $\mathbb{Q}$ is dense in $\mathbb{R}$. Also, from the Weierstrass approximation theorem of analysis we can observe that the set of polynomials is dense in the set $C[a,b]$ of continuous maps with the norm $\| \cdot \|_\infty$. As in Example 2.4.14, we can observe that the closure of an open ball in a metric space need not be the closed ball.

Proposition 2.4.39. *Let V be a normed space, and let $x \in V$. Then $\operatorname{Cl} B(x,r) = D(x,r)$ for all $r > 0$.*

Proof. Since $B(x,r) \subseteq D(x,r)$, $\operatorname{Cl} B(x,r) \subseteq D(x,r)$. For the converse, let $y \notin \operatorname{Cl} B(x,r)$. Since $\operatorname{Cl} B(x,r)$ is closed, $X \setminus \operatorname{Cl} B(x,r)$ is open. Then there exists $\epsilon > 0$ such that

$$B(y,\epsilon) \subseteq X \setminus \operatorname{Cl} B(x,r) \subseteq X \setminus B(x,r).$$

Then $\|x - y\| \geq r$. If $\|x - y\| > r$, then $y \notin D(x,r)$. Now suppose that $\|x - y\| = r$. Let

$$z = y - \frac{\epsilon}{2} \frac{y-x}{\|x-y\|}.$$

Note that $\|z - y\| = \frac{\epsilon}{2} < \epsilon$ and $\|z - x\| = r - \frac{\epsilon}{2} < r$. This implies that $z \in B(y,\epsilon) \cap B(x,r)$. This is a contradiction. Hence $D(x,r) \subseteq \operatorname{Cl} B(x,r)$. $\square$

Note that A is closed if and only if $\operatorname{Cl} A = A$. Suppose that $A \subseteq B$. Then $A \subseteq B \subseteq \operatorname{Cl} B$. By the definition of the closure, $\operatorname{Cl} A \subseteq \operatorname{Cl} B$.

Proposition 2.4.40. *Let $\{A_\alpha \mid \alpha \in \mathcal{I}\}$ be a family of sets in a metric space. Then*
(i) $\bigcup_{\alpha \in \mathcal{I}} \operatorname{Cl} A_\alpha \subseteq \operatorname{Cl}(\bigcup_{\alpha \in \mathcal{I}} A_\alpha)$;
(ii) *if $\mathcal{I}$ is finite, then $\bigcup_{\alpha \in \mathcal{I}} \operatorname{Cl} A_\alpha = \operatorname{Cl}(\bigcup_{\alpha \in \mathcal{I}} A_\alpha)$;*
(iii) $\operatorname{Cl}(\bigcap_{\alpha \in \mathcal{I}} A_\alpha) \subseteq \bigcap_{\alpha \in \mathcal{I}} \operatorname{Cl} A_\alpha.$

Proof. We will only prove (ii). The others are left as exercises.
Without loss of generality, assume that $\mathcal{I} = \{1, \ldots, n\}$. Let $F = \bigcup_{\alpha=1}^{n} \operatorname{Cl} A_\alpha$. Note that F is closed. Since $A_\alpha \subseteq \operatorname{Cl} A_\alpha$,

$$\bigcup_{\alpha=1}^{n} A_\alpha \subseteq \bigcup_{\alpha=1}^{n} \operatorname{Cl} A_\alpha = F.$$

Since for any set B, $\operatorname{Cl} B$ is the smallest closed set containing B,

$$\operatorname{Cl}\left(\bigcup_{\alpha=1}^{n} A_\alpha \right) \subseteq \bigcup_{\alpha=1}^{n} \operatorname{Cl} A_\alpha.$$

The equality holds because of (i). $\square$

Proposition 2.4.41. *Let A be a set in a metric space X. Then $x \in \mathrm{Cl}\,A$ if and only if for every open set U containing x, $U \cap A \neq \emptyset$.*

Proof. Let $x \in \mathrm{Cl}\,A$. On the contrary, suppose that there is an open set U containing x such that $U \cap A = \emptyset$. Then $A \subseteq X \setminus U$. Since $X \setminus U$ is closed, $\mathrm{Cl}\,A \subseteq X \setminus U$. This implies that $x \in \mathrm{Cl}\,A \subseteq X \setminus U$. This is a contradiction. Conversely, suppose that $x \notin \mathrm{Cl}\,A$. This implies that $x \in X \setminus \mathrm{Cl}\,A$. Let $U = X \setminus \mathrm{Cl}\,A$. Note that U is open and $U = X \setminus \mathrm{Cl}\,A \subseteq X \setminus A$. Therefore $U \cap A = \emptyset$. $\qquad\square$

Proposition 2.4.42. *Let A be a set in a metric space X. Then $\mathrm{Bd}\,A = \mathrm{Cl}\,A \cap \mathrm{Cl}(X \setminus A)$.*

Proof. Note that $x \in \mathrm{Bd}\,A$ if and only if x is neither an interior nor an exterior point of A. This equivalently means that for each open set U containing x, we have

$$U \nsubseteq A \text{ and } U \nsubseteq X \setminus A$$
$$\Leftrightarrow U \cap A \neq \emptyset \text{ and } U \cap (X \setminus A) \neq \emptyset.$$

Therefore by Proposition 2.4.41 we get $\mathrm{Bd}\,A = \mathrm{Cl}\,A \cap \mathrm{Cl}(X \setminus A)$. $\qquad\square$

Proposition 2.4.43. *Let Y be a subspace of a metric space X, and let $A \subseteq Y$. Then $\mathrm{Cl}_Y A = \mathrm{Cl}_X A \cap Y$.*

Proof. Note that

$$\mathrm{Cl}_Y A = \bigcap_\alpha \{F_\alpha \mid F_\alpha \text{ is closed in } Y \text{ and } A \subseteq F_\alpha\}$$
$$= \bigcap_\alpha \{K_\alpha \cap Y \mid K_\alpha \text{ is closed in } X \text{ and } A \subseteq K_\alpha \cap Y\}$$
$$= \bigcap_\alpha \{K_\alpha \mid K_\alpha \text{ is closed in } X \text{ and } A \subseteq K_\alpha\} \cap Y$$
$$= \mathrm{Cl}_X A \cap Y. \qquad\qquad\square$$

Proposition 2.4.44. *Let A be a set in a metric space X. Then*
(i) $\mathrm{Cl}(X \setminus A) = X \setminus \mathrm{Int}\,A$;
(ii) $\mathrm{Int}(X \setminus A) = X \setminus \mathrm{Cl}\,A$.

Proof. We will only prove (i). The proof of (ii) follows the same lines.
Note that

$$X \setminus \mathrm{Int}\,A = X \setminus \bigcup_\alpha \{U_\alpha \mid U_\alpha \text{ is open in } X \text{ and } U_\alpha \subseteq A\}$$
$$= \bigcap_\alpha \{X \setminus U_\alpha \mid X \setminus U_\alpha \text{ is closed in } X \text{ and } X \setminus A \subseteq X \setminus U_\alpha\}$$
$$= \mathrm{Cl}(X \setminus A). \qquad\qquad\square$$

Proposition 2.4.45. *Let A be a set in a metric space X. Then* $\operatorname{Cl} A = \operatorname{Int} A \cup \operatorname{Bd} A$.

Proof. By Propositions 2.4.42 and 2.4.44 we have

$$
\begin{aligned}
\operatorname{Bd} A &= \operatorname{Cl} A \cap \operatorname{Cl}(X \setminus A) \\
&= \operatorname{Cl} A \cap (X \setminus \operatorname{Int} A) \\
&= \operatorname{Cl} A \setminus \operatorname{Int} A.
\end{aligned}
$$

Therefore $\operatorname{Cl} A = \operatorname{Int} A \cup \operatorname{Bd} A$. □

Definition 2.4.46. Let A be a set in a metric space X. A point $x \in X$ is called a limit point of A if for each $r > 0$,

$$
(B(x, r) \setminus \{x\}) \cap A \neq \emptyset.
$$

In other words, $x \in X$ is a limit point of A if each open ball around x contains a point A other than x itself. We denote the set of all limit points of A by $D(A)$. The set $D(A)$ is called the derived set of A.

Example 2.4.47. Consider the open interval (a, b) in the real line. Observe that if $x \in [a, b]$, then every open ball $B(x, r)$ around x contains a point of (a, b) other than x itself. Let $x \in \mathbb{R} \setminus [a, b]$. Then either $x < a$ or $x > b$. If $x < a$, then the open ball $B(x, a - x)$ contains no point of (a, b). If $x > b$, then the open ball $B(x, x - b)$ contains no point of (a, b). This shows that $D((a, b)) = [a, b]$. We can similarly show that $D([a, b]) = D([a, b)) = D((a, b]) = [a, b]$.

Example 2.4.48. Since every open interval around any real number contains rational as well as irrational numbers, $D(\mathbb{Q}) = D(\mathbb{R} \setminus \mathbb{Q}) = \mathbb{R}$.

Example 2.4.49. Consider the set $A = \{\frac{1}{n} \mid n \in \mathbb{N}\}$ in the real line. We will prove that $D(A) = \{0\}$. Let $r > 0$. By the Archimedean property there exists $k \in \mathbb{N}$ such that $\frac{1}{k} < r$. This implies that the open interval $(-r, r)$ contains a point of A. This shows that $0 \in D(A)$. Now suppose that $x \neq 0$. Then we have the following possibilities:
(i) $x \in (\frac{1}{k}, \frac{1}{k+1})$ for some $k \in A$,
(ii) $x \in A$,
(iii) either $x < 0$ or $x > 1$.

We can find a suitable open interval in each of the above cases that does not contain a point of A other than possibly x. Hence $D(A) = \{0\}$.

Example 2.4.50. Consider the set $\mathbb{Z}$ of integers in the real line. Let $x \in \mathbb{R}$. Since $(x - \frac{1}{2}, x + \frac{1}{2})$ contains at most one point of $\mathbb{Z}$, possibly x itself, $D(\mathbb{Z}) = \emptyset$.

Example 2.4.51. In the discrete metric space X, $D(A) = \emptyset$ for every set A in X.

Proposition 2.4.52. *Let A and B be sets in a metric space X. Then*

(i) *if $A \subseteq B$, then $D(A) \subseteq D(B)$;*
(ii) $D(A \cup B) = D(A) \cup D(B)$.

Proof. We can easily observe (i). We will prove (ii). Since $A, B \subseteq A \cup B$, $D(A), D(B) \subseteq D(A \cup B)$. This implies that $D(A) \cup D(B) \subseteq D(A \cup B)$.

Now suppose that $x \notin D(A) \cup D(B)$. This implies that $x \notin D(A)$ and $x \notin D(B)$. Therefore there are open balls $B(x, r_1)$ and $B(x, r_2)$ such that $B(x, r_i) \cap A \subseteq \{x\}$, $i = 1, 2$. Let $r = \min\{r_1, r_2\}$. Then

$$B(x, r) \cap (A \cup B) = \big(B(x, r) \cap A\big) \cup \big(B(x, r) \cap B\big)$$
$$\subseteq \{x\}.$$

This shows that $x \notin D(A \cup B)$. Hence $D(A \cup B) = D(A) \cup D(B)$. ☐

Proposition 2.4.53. *Let A be a set in a metric space X, and let $x \in X$ be a limit point of A. Then every open ball around x contains infinitely many points of A other than x.*

Proof. Suppose there is a positive real number r such that the open ball $B(x, r)$ around x contains finitely many points of A other than x. Let these points be $a_1, a_2, \ldots, a_n$. Note that $r_i = d(x, a_i) > 0$ for all $1 \le i \le n$. Let

$$s = \min\{r_1, r_2, \ldots, r_n\}.$$

Then the open ball $B(x, s)$ does not contain any point of A other than x. This implies that x is not a limit point of A, a contradiction. ☐

Proposition 2.4.54. *A set A in a metric space X is closed if and only if $D(A) \subseteq A$.*

Proof. Suppose that A is closed in X. Let $x \in D(A)$. On the contrary, suppose that $x \notin A$. Then $x \in X \setminus A$. Since $X \setminus A$ is open, there is an open ball $B(x, r)$ around x contained in $X \setminus A$. This implies that $B(x, r)$ contains no point of A, a contradiction. Hence $x \in A$.

Conversely, suppose that $D(A) \subseteq A$. Let $x \in X \setminus A$. Then $x \notin D(A)$. This implies that there is an open ball $B(x, r)$ that does not contain any point of A other than x. Since $x \notin A$, $B(x, r) \subseteq X \setminus A$. This shows that $X \setminus A$ is open in X. Hence A is closed in X. ☐

Proposition 2.4.55. *A set A in a metric space X is closed if and only if $\operatorname{Bd} A \subseteq A$.*

Proof. Suppose that A is closed in X. Let $x \in \operatorname{Bd} A$. Then $B(x, r) \cap A \ne \emptyset$ for all $r > 0$. On the contrary, suppose that $x \notin A$. This implies that $(B(x, r) \setminus \{x\}) \cap A \ne \emptyset$. This shows that $x \in D(A)$. By Proposition 2.4.54, $x \in A$. This is a contradiction.

Conversely, suppose that $\operatorname{Bd} A \subseteq A$. Let $x \in D(A)$. Then $(B(x, r) \setminus \{x\}) \cap A \ne \emptyset$. On the contrary, suppose that $x \notin A$. Then $x \in X \setminus A$. This implies that $B(x, r) \cap (X \setminus A) \ne \emptyset$. This shows that $x \in \operatorname{Bd} A$. Hence $x \in A$, a contradiction. ☐

Proposition 2.4.56. *Let A be a set in a metric space X. Then $D(A)$ is a closed set in X.*

Proof. To prove this, we will prove that $D(D(A)) \subseteq D(A)$. Let $x \in D(D(A))$. Then for each $r > 0$,

$$(B(x,r) \setminus \{x\}) \cap D(A) \neq \emptyset.$$

Let $y \in (B(x,r) \setminus \{x\}) \cap D(A)$. Observe that $B(x,r) \setminus \{x\}$ is an open set. Then there is an open ball $B(y,s)$ contained in $B(x,r) \setminus \{x\}$. Since $y \in D(A)$,

$$(B(y,s) \setminus \{y\}) \cap A \neq \emptyset.$$

Let $z \in (B(y,s) \setminus \{y\}) \cap A$. Then

$$z \in B(y,s) \setminus \{y\} \subseteq B(x,r) \setminus \{x\}.$$

This implies that $z \in (B(x,r) \setminus \{x\}) \cap A$. This shows that $x \in D(A)$. □

Proposition 2.4.57. *Let A be a set in a metric space X. Then* $\operatorname{Cl} A = A \cup D(A)$.

Proof. Since $\operatorname{Cl} A$ is closed, by Proposition 2.4.54 $D(\operatorname{Cl} A) \subseteq \operatorname{Cl} A$. Since $A \subseteq \operatorname{Cl} A$, $D(A) \subseteq D(\operatorname{Cl} A)$. This implies that $A \cup D(A) \subseteq \operatorname{Cl} A$. To prove the equality, we will prove that $A \cup D(A)$ is the smallest closed set containing A. For this, let $x \in X \setminus (A \cup D(A))$. This implies that $x \notin A$ and $x \notin D(A)$. Since $D(A)$ is closed, we can find an open ball $B(x,r)$ contained in $X \setminus A$ such that no point of $B(x,r)$ is in $D(A)$. Therefore $B(x,r) \subseteq X \setminus (A \cup D(A))$. This implies that x is an interior point of $X \setminus (A \cup D(A))$. This shows that $X \setminus (A \cup D(A))$ is open set in X. Hence $A \cup D(A)$ is closed.

Now suppose that F is a closed set in X containing A. Then $D(A) \subseteq D(F)$. Since F is closed, $D(F) \subseteq F$. This shows that $A \cup D(A) \subseteq F$. Thus $A \cup D(A)$ is the smallest closed set containing A. □

By Proposition 2.4.57 we observe that if A is dense in a metric space X, then the points of X can be approximated by points of A.

Definition 2.4.58. A metric space X is called separable if it has a countable dense set.

Example 2.4.59. By the rational density theorem the real line is separable.

Example 2.4.60. The Euclidean space $\mathbb{R}^n$ is separable, since $\mathbb{Q}^n$ is dense in $\mathbb{R}^n$.

Example 2.4.61. The discrete metric space is separable if and only if it is countable.

Example 2.4.62. The space ℓ_2 is separable. We may look forward to the set $\{(x_n) \mid x_n \in \mathbb{Q}\}$ as a desired one, but this is an uncountable set.

For each $n \in \mathbb{N}$, let us consider the set

$$A_n = \{(y_m) \mid y_1, \ldots, y_n \in \mathbb{Q}, \text{ and } y_m = 0 \text{ for all } m > n\}.$$

Let $A = \bigcup_{n=1}^{\infty} A_n$. Since each A_n is countable, A is countable. We claim that A is dense in ℓ_2. Let $\epsilon > 0$ and $(x_m) \in \ell_2$. Then there is $k \in \mathbb{N}$ such that

$$\sum_{m=k+1}^{\infty} x_m^2 < \frac{\epsilon^2}{2}.$$

For each $1 \le m \le k$, choose a rational number y_m such that

$$|x_m - y_m|^2 < \frac{\epsilon^2}{2k}.$$

Define a sequence $y = (y_m)$ such that $y_m = 0$ for all $m > k$. Then $y \in A_k$. Now

$$\|x - y\|_2^2 = \sum_{m=1}^{\infty} |x_m - y_m|^2$$

$$= \sum_{m=1}^{k} |x_m - y_m|^2 + \sum_{m=k+1}^{\infty} |x_m - y_m|^2$$

$$< k\frac{\epsilon^2}{2k} + \frac{\epsilon^2}{2}$$

$$= \epsilon^2.$$

This implies that A is dense in ℓ_2.

Example 2.4.63. Consider the space $C[a, b]$ of real-valued functions on $[a, b]$ with the supremum norm. By the Weierstrass approximation theorem each continuous function can be approximated by a real polynomial

$$p(X) = a_0 + a_1 X + \cdots + a_n X^n, \quad a_i \in \mathbb{R}.$$

Since each real number can be approximated by a rational number, each function in $C[a, b]$ can be approximated by a polynomial with rational coefficients. We can observe that the set of all polynomials with rational coefficients is countable. This shows that $C[a, b]$ is separable.

Example 2.4.64. Consider the space $B[0, 1]$ of bounded real-valued functions on $[0, 1]$ with the norm

$$\|f\|_\infty = \sup\{|f(x)| \mid x \in [0, 1]\}.$$

We claim that $B[0, 1]$ is not separable. For each $x \in [0, 1]$, consider the map $\chi_x : [0, 1] \to \mathbb{R}$ defined by

$$\chi_x(y) = \begin{cases} 1 & \text{if } y = x, \\ 0 & \text{if } y \ne x. \end{cases}$$

Then $\chi_x \in B[0,1]$. Let A be a dense set in $B[0,1]$. Then for each $x \in [0,1]$, there is a function $f_x \in A$ such that

$$\|f_x - \chi_x\|_\infty < \frac{1}{2}.$$

Let $x, y \in [0,1]$ be such that $x \neq y$. Then

$$\|\chi_x - \chi_y\|_\infty = 1.$$

If $f_x = f_y$, then by the triangle inequality we have

$$\begin{aligned}
1 &= \|\chi_x - \chi_y\|_\infty \\
&\leq \|\chi_x - f_x\|_\infty + \|f_y - \chi_y\|_\infty \\
&< \frac{1}{2} + \frac{1}{2} \\
&= 1,
\end{aligned}$$

a contradiction. Therefore $f_x \neq f_y$ for $x \neq y$. This shows that

$$\{f_x \mid x \in [0,1]\} \subseteq A.$$

Hence A is uncountable.

2.5 Distance between sets

In elementary geometry of two or three dimensions, we study the distance between a point and a line or a point and a plane as the shortest distance. With the same motivation, we can define the distance between a point and a set in a metric space.

Definition 2.5.1. Let A be a nonempty set in a metric space X, and let $x \in X$. Then the real number

$$d(x, A) = \inf\{d(x, y) \mid y \in A\}$$

is called the distance of the point x from the set A.

Note that $d(x, A) \leq d(x, y)$ for all $y \in A$ and for each real number $a > d(x, A)$, there exists an element $z \in A$ such that $d(x, z) < a$. We can easily observe that $d(x, A) \in [0, \infty)$ and if B is a set in X such that $A \subseteq B$, then $d(x, A) \leq d(x, B)$.

Example 2.5.2. If $x \in A$, then $d(x, A) = 0$. The value $d(x, A)$ may be zero for $x \notin A$. For example, consider the set $A = \{\frac{1}{n} \mid n \in \mathbb{N}\}$ in the real line. Note that $0 \notin A$ and $d(0, A) = \inf\{d(0, \frac{1}{n}) \mid n \in \mathbb{N}\} = 0$.

Proposition 2.5.3. *Let A be a nonempty set in a metric space X, and let $x, y \in X$. Then*
(i) $d(x, A) = 0$ *if and only if* $x \in \operatorname{Cl} A$,
(ii) $d(x, A) \leq d(x, y) + d(y, A)$.

Proof. (i) Note that

$$x \in \operatorname{Cl} A \Leftrightarrow B(x, r) \cap A \neq \emptyset \text{ for all } r > 0$$
$$\Leftrightarrow \text{for all } r > 0, \text{ there exists } a \in A \text{ such that } d(x, a) < r$$
$$\Leftrightarrow \inf\{d(x, y) \mid y \in A\} = 0.$$

(ii) Let $z \in A$. Then

$$d(x, z) \leq d(x, y) + d(y, z).$$

This implies that

$$d(y, z) \geq d(x, z) - d(x, y)$$
$$\geq d(x, A) - d(x, y).$$

Since $z \in A$ is arbitrary, $d(y, A) \geq d(x, A) - d(x, y)$. In other words, $d(x, A) \leq d(x, y) + d(y, A)$. $\qquad\square$

Proposition 2.5.4. *Let F_1 and F_2 be disjoint closed subsets of a metric space X. Then there exist disjoint open sets U_1 and U_2 of X containing F_1 and F_2, respectively.*

Proof. For each $x \in F_1$, define

$$r_x = \frac{d(x, F_2)}{2},$$

and for each $y \in F_2$, define

$$s_y = \frac{d(y, F_1)}{2}.$$

Let

$$U_1 = \bigcup_{x \in F_1} B(x, r_x)$$

and

$$U_2 = \bigcup_{y \in F_2} B(x, s_y).$$

Note that U_1 and U_2 are open sets containing F_1 and F_2, respectively. We will show that U_1 and U_2 are disjoint.

On the contrary, suppose that $z \in U_1 \cap U_2$. Then for some $x \in F_1$ and $y \in F_2$, we have

$$d(x,z) < r_x$$

and

$$d(y,z) < s_y.$$

Without loss of generality, assume that $s_y \leq r_x$. Then

$$d(x,y) \leq d(x,z) + d(z,y)$$
$$< r_x + s_y$$
$$\leq 2r_x = d(x, F_2).$$

This is a contradiction, since $d(x, F_2) \leq d(x,y)$. $\qquad\square$

Proposition 2.5.5. *Let W be a subspace of a finite-dimensional inner product space V, and let $\{x_1, x_2, \ldots, x_m\}$ be an orthonormal basis of W. Let $x \in V$. Then*

$$d(x, W) = \sqrt{\|x\|^2 - \sum_{k=1}^{m} |\langle x, x_k \rangle|^2}.$$

Proof. Note that $z = \sum_{k=1}^{m} \langle x, x_k \rangle x_k \in W$ and

$$d(x,z)^2 = \left\| x - \sum_{k=1}^{m} \langle x, x_i \rangle x_k \right\|^2 = \|x\|^2 - \sum_{k=1}^{m} |\langle x, x_i \rangle|^2.$$

Let $y = \sum_{k=1}^{m} a_k x_k$ in W. Then

$$d(x,y)^2 = \|x - y\|^2$$
$$= \|x\|^2 + \sum_{k=1}^{m} |a_k|^2 - \sum_{k=1}^{m} \overline{a_k} \langle x, x_k \rangle - \sum_{k=1}^{m} a_k \overline{\langle x, x_k \rangle}.$$

Note that $w + \overline{w} \leq 2|w|$ for all $w \in \mathbb{F}$ ($\mathbb{F} = \mathbb{R}$ or $\mathbb{C}$). Then by the Cauchy–Schwarz inequality we get

$$\sum_{k=1}^{m} \overline{a_k} \langle x, x_k \rangle + \sum_{k=1}^{m} a_k \overline{\langle x, x_k \rangle} \leq 2 \left| \sum_{k=1}^{m} \overline{a_k} \langle x, x_k \rangle \right|$$

$$\leq 2 \sqrt{\sum_{k=1}^{m} |a_k|^2} \sqrt{\sum_{k=1}^{m} |\langle x, x_k \rangle|^2}.$$

Therefore

$$d(x,y)^2 = \|x - y\|^2$$

$$\geq \left\| x - \sum_{k=1}^{m} \langle x, x_k \rangle x_k \right\|^2 = \|x\|^2 - \sum_{k=1}^{m} |\langle x, x_i \rangle|^2$$

$$= d(x,z)^2.$$

Since $y \in W$ is arbitrary,

$$d(x, W) = \sqrt{\|x\|^2 - \sum_{i=1}^{m} |\langle x, x_k \rangle|^2}.$$ $\square$

Remark 2.5.6. The point z in the proof of Proposition 2.5.5 such that $d(x, W) = d(x, z)$ is unique. Indeed, suppose that $z_1 = \sum_{k=1}^{m} a_k x_k \in W$ is such that $d(x, z_1) = d(x, W) = d(x, z)$. This implies that

$$\left\| x - \sum_{k=1}^{m} a_k x_k \right\|^2 = \|x\|^2 - \sum_{i=1}^{m} |\langle x, x_k \rangle|^2.$$

Therefore we get

$$\sum_{k=1}^{m} |a_k|^2 + \sum_{k=1}^{m} |\langle x, x_k \rangle|^2 = \sum_{k=1}^{m} \overline{a_k} \langle x, x_k \rangle + \sum_{k=1}^{m} a_k \overline{\langle x, x_k \rangle}.$$

Let $w_1 = (a_1, \ldots, a_m)$ and $w_2 = (\langle x, x_1 \rangle, \ldots, \langle x, x_m \rangle)$ belong to $\mathbb{F}^m$. If we consider the standard inner product on $\mathbb{F}^n$, then the above equation can be rewritten as

$$\|w_1\|^2 + \|w_2\|^2 = \langle w_1, w_2 \rangle + \langle w_2, w_1 \rangle.$$

In other words,

$$\langle w_1, w_1 \rangle + \langle w_2, w_2 \rangle = \langle w_1, w_2 \rangle + \langle w_2, w_1 \rangle.$$

This implies that $\langle w_1 - w_2, w_1 - w_2 \rangle = 0$. Hence $w_1 = w_2$. Therefore $a_i = \langle x, x_i \rangle$ for all i. This shows that $z = z_1$.

Corollary 2.5.7. *Let W be a subspace of a finite-dimensional inner product space V, and let $\{x_1, x_2, \ldots, x_m\}$ be an orthonormal basis of W. Let $x, y \in V$. Then*

$$d(x, y + W) = \sqrt{\|x - y\|^2 - \sum_{k=1}^{m} |\langle x, x_k \rangle|^2}.$$

Proof. Note that

$$d(x, y + W) = \inf\{d(x, y + w) \mid w \in W\}$$
$$= \inf\{\|x - y - w\| \mid w \in W\}$$
$$= d(x - y, W)$$
$$= \sqrt{\|x - y\|^2 - \sum_{k=1}^{m} |\langle x, x_k \rangle|^2}.$$ $\qquad\square$

Proposition 2.5.8. *Let V be a normed space over a field $\mathbb{F}$, and let $f \in B(V, \mathbb{F})$ be such that $f \neq 0$. Let $W = \{a \in V \mid f(a) = 0\}$. Then*

$$d(x, W) = \frac{|f(x)|}{\|f\|}.$$

Proof. Let $y \in W$. Then $|f(x)| = |f(x - y)| \leq \|f\|\|x - y\|$. Since $\|f\| \neq 0$, for all $y \in W$, we have

$$\frac{|f(x)|}{\|f\|} \leq \|x - y\|.$$

This implies that

$$\frac{|f(x)|}{\|f\|} \leq \inf\{\|x - y\| \mid y \in W\} = d(x, W).$$

Recall that

$$\|f\| = \sup\{|f(x)| \mid \|x\| \leq 1\}.$$

This implies that for $0 < \epsilon < \|f\|$, there is $z \in V$ with $\|z\| \leq 1$ such that

$$0 < \epsilon < \|f\| < |f(z)| < \|f\|.$$

Let $w = x - \frac{f(x)}{f(z)} z$. Then $w \in W$. This implies that

$$d(x, W) \leq \|x - w\| = \frac{|f(x)|}{|f(z)|} \|z\|$$
$$\leq \frac{|f(x)|}{|f(z)|}$$
$$\leq \frac{|f(x)|}{\|f\|}.$$

Hence $d(x, W) = \frac{|f(x)|}{\|f\|}$. $\qquad\square$

Corollary 2.5.9. *Let V be a normed space over a field $\mathbb{F}$, and let $f \in B(V, \mathbb{F})$ be such that $f \neq 0$. Let $A = \{x \in V \mid f(x) = a\}$, where $a \in \mathbb{F}$. Then*

$$d(x, A) = \frac{|f(x) - a|}{\|f\|}.$$

Proof. Since $f \neq 0, f(u) \neq 0$ for some $u \in V$. Let $z = \frac{a}{f(u)} u$. Then $z \in A$ and $A = z + W$, where $W = \{x \in V \mid f(x) = 0\}$. Now by Proposition 2.5.8 we have

$$
\begin{aligned}
d(x, A) &= \inf\{\|x - y\| \mid y \in A\} \\
&= \inf\{\|x - z - w\| \mid w \in W\} \\
&= d(x - z, W) \\
&= \frac{|f(x - z)|}{\|f\|} \\
&= \frac{|f(x) - a|}{\|f\|}.
\end{aligned}
$$

$\square$

Let us verify that the above formula is the same that we have studied in the elementary geometry. Consider a plane $ax + by + cz + d = 0$ in $\mathbb{R}^3$, where $(a, b, c) \neq (0, 0, 0)$. Consider the subspace $W = \{(x, y, z) \mid ax + by + cz = 0\}$ of $\mathbb{R}^3$. Let $(x_1, y_1, z_1) \in R^3$ be such that $ax_1 + by_1 + cz_1 + d = 0$. Then $A = W + (x_1, y_1, z_1) = \{(x, y, z) \mid ax + by + cz + d = 0\}$. Define the map $f : \mathbb{R}^3 \to \mathbb{R}$ by $f(x, y, z) = ax + by + cz$. Check that $\|f\| = \sqrt{a^2 + b^2 + c^2}$. Then by Corollary 2.5.9,

$$d((u, v, w), A) = \frac{au + bv + cw + d}{\sqrt{a^2 + b^2 + c^2}}.$$

Now we define the distance between sets in a metric space.

Definition 2.5.10. Let A and B be nonempty sets in a metric space X. Then the real number

$$d(A, B) = \inf\{d(x, y) \mid x \in A, y \in B\}$$

is called the distance between the sets A and B.

We can easily observe that $d(A, B) \in [0, \infty)$ and $d(A, B) = d(B, A)$ for all nonempty sets A and B in a metric space. We can also observe that $d(A_1, B_1) \leq d(A, B)$ for all nonempty subsets A_1 and B_1 of A and B, respectively

Proposition 2.5.11. *Let A, B, and C be nonempty sets in a metric space X. Then*
(i) $d(A, B) = \inf\{d(x, B) \mid x \in A\}$,
(ii) $d(A \cup B, C) = \min\{d(A, C), d(B, C)\}$,
(iii) $d(A, B) \leq d(x, A) + d(x, B)$ *for all $x \in X$.*

Proof. (i) Note that

$$\{d(x, y) \mid x \in A, y \in B\} = \bigcup_{x \in A} \{d(x, y) \mid y \in B\}.$$

This implies that

$$d(A,B) = \inf\{d(x,y) \mid x \in A, y \in B\}$$
$$= \inf\{\inf\{d(x,y) \mid y \in B\} \mid x \in A\}$$
$$= \inf\{d(x,B) \mid x \in A\}.$$

(ii) We can easily observe the following:

$$\{d(x,C) \mid x \in A \cup B\} = \{d(x,C) \mid x \in A\} \cup \{d(x,C) \mid x \in B\}.$$

Therefore (ii) follows from (i).

(iii) Let $\epsilon > 0$. Then there are $a \in A$ and $b \in B$ such that

$$d(x,a) < d(x,A) + \frac{\epsilon}{2}$$

and

$$d(x,b) < d(x,B) + \frac{\epsilon}{2}.$$

Then

$$d(A,B) \leq d(a,b)$$
$$\leq d(a,x) + d(x,b)$$
$$< d(x,A) + d(x,B) + \epsilon.$$

Since $\epsilon > 0$ is arbitrary,

$$d(A,B) \leq d(x,A) + d(x,B). \qquad \square$$

The distance between sets is not a metric on the set of sets in a metric space. It may not satisfy condition (ii) of Definition 2.1.1. For example, consider $A = (0,1)$ and $B = (1,2)$ in the real line. Note that $d(A,B) = 0$. Similarly, it may not satisfy the triangle inequality. For example, consider $A = (0,1)$, $B = (4,5)$, and $C = (\frac{3}{2},2) \cup (\frac{5}{2},\frac{7}{2})$. Then $d(A,B) = 3$, $d(A,C) = \frac{1}{2}$, and $d(B,C) = \frac{1}{2}$.

Now we define the distance between bounded sets of a metric space that satisfies most conditions of the metric.

Definition 2.5.12. Let A and B be nonempty bounded sets in a metric space X. Then the real number

$$d_H(A,B) = \max\{\sup\{d(x,B) \mid x \in A\}, \sup\{d(y,A) \mid y \in B\}\}$$

is called the Hausdorff distance between the sets A and B.

We can easily observe that $d_H(A, B) \in [0, \infty)$ and $d_H(A, B) = d_H(B, A)$.

Proposition 2.5.13. *Let A, B, and C be nonempty bounded sets in a metric space X. Then*
(i) $d_H(A, B) = 0$ *if and only if* $\operatorname{Cl} A = \operatorname{Cl} B$,
(ii) $d_H(A, B) \le d_H(A, C) + d_H(C, B)$.

Proof. By Proposition 2.5.3(i) we can observe that $d_H(A, B) = 0$ if and only if $A \subseteq \operatorname{Cl} B$ and $B \subseteq \operatorname{Cl} A$. This proves (i). We will now prove (ii). Let $x \in A$. Then by Proposition 2.5.3(ii), for any $y \in C$, we have

$$d(x, B) \le d(x, y) + d(y, B)$$
$$\le d(x, y) + d_H(C, B).$$

This implies that

$$d(x, B) - d_H(C, B) \le d(x, y)$$
$$\le d(x, C)$$
$$\le d_H(A, C).$$

Therefore

$$d(x, B) \le d_H(A, C) + d_H(C, B).$$

Similarly, for $z \in B$, we have

$$d(z, A) \le d_H(A, C) + d_H(C, B).$$

Hence

$$d_H(A, B) \le d_H(A, C) + d_H(C, B). \qquad \square$$

By Proposition 2.5.13(i) we observe that $d_H(A, A) = 0$ for all nonempty bounded sets A in X. We can also observe that d_H satisfies all the conditions of a metric on the set of nonempty bounded sets in X except that $d_H(A, B)$ may be zero without $A = B$.

Proposition 2.5.14. *The Hausdorff distance d_H is a metric on the set of all nonempty closed bounded sets of a metric space X.*

Proof. We only have to prove that $d_H(A, B) = 0$ implies $A = B$. For this, suppose A and B are nonempty closed bounded sets of a metric space X such that $A \ne B$. We can suppose that $A \setminus B \ne \emptyset$. Let $a \in A \setminus B$. Since B is closed, $X \setminus B$ is open. Then there is a positive real number r such that $B(a, r) \subseteq X \setminus B$. This implies that $d(a, B) \ge r$. Hence

$$d_H(A, B) \ge d(a, B) \ge r > 0. \qquad \square$$

We now prove an equivalent formulation of the Hausdorff distance. For a nonempty set A in a metric space X, let

$$B(A,r) = \{y \in X \mid d(x,A) < r\} \quad \text{and}$$
$$D(A,r) = \{y \in X \mid d(x,A) \leq r\}.$$

Proposition 2.5.15. *Let A and B be nonempty bounded sets in a metric space X. Then*

$$d_H(A,B) = \inf\{r \in [0,\infty) \mid A \subseteq B(A,r) \text{ and } B \subseteq B(B,r)\}$$
$$= \inf\{r \in [0,\infty) \mid A \subseteq D(A,r) \text{ and } B \subseteq D(B,r)\}.$$

Proof. Let

$$d_1 = \inf\{r \in [0,\infty) \mid A \subseteq B(A,r) \text{ and } B \subseteq B(B,r)\}$$

and

$$d_2 = \inf\{r \in [0,\infty) \mid A \subseteq D(A,r) \text{ and } B \subseteq D(B,r)\}.$$

Note that for nonempty sets C and D in X, if $C \subseteq B(D,r)$, then $C \subseteq D(D,r)$. This implies that $d_2 \leq d_1$. Also, note that $d(a,B) \leq d_H(A,B)$ and $d(b,A) \leq d_H(A,B)$ for all $a \in A$ and $b \in B$. This shows that $A \subseteq B(B,r)$ and $B \subseteq B(A,r)$ for all $r > d_H(A,B)$. Therefore $d_1 \leq r$ for all $r > d_H(A,B)$. Hence $d_1 \leq d_H(A,B)$.

Now note that $A \subseteq D(B,r)$ and $B \subseteq D(A,r)$ for all $r > d_2$. Therefore, for all $a \in A$ and $b \in B$, we have $d(a,B) \leq r$ and $d(b,A) \leq r$. Hence $d_H(A,B) \leq r$ for all $r > d_2$. This implies that $d_H(A,B) \leq d_2$. Thus

$$d_H(A,B) \leq d_2 \leq d_1 \leq d_H(A,B).$$

This shows that $d_H(A,B) = d_1 = d_2$. $\qquad\square$

Let (X,d_1) and (Y,d_2) be bounded metric spaces. We would like to ask "What is the distance between X and Y?" At the first sight, this question may seem to be absurd, but if X and Y are subspaces of some larger space Z, then we can think of the Hausdorff distance between X and Y. Let us consider the following example.

Example 2.5.16. Let (X,d_1) and (Y,d_2) be bounded metric spaces such that $X \cap Y = \emptyset$. Let $Z = X \sqcup Y$. Define $d : Z \times Z \to \mathbb{R}$ by

$$d(x,y) = \begin{cases} d_1(x,y) & \text{if } x,y \in X, \\ d_2(x,y) & \text{if } x,y \in Y, \\ m & \text{otherwise}, \end{cases}$$

where $m = \max\{\operatorname{diam}(X), \operatorname{diam}(Y)\}$. We can easily observe that d is nonnegative and symmetric and that d is zero only on the diagonal of Z. We now observe that d satisfies the triangle inequality. We will prove it in two cases, and the remaining cases are left as exercises. Suppose $x, y \in X$ and $z \in Y$. Then

$$d(x,y) = d_1(x,y) \le m + m = d(x,z) + d(z,y).$$

Suppose that $x, z \in X$ and $y \in Y$. Then

$$d(x,y) = m \le m + d(x,z) = d(z,y) + d(x,z).$$

The other cases can be similarly proved.

Definition 2.5.17. Let (X, d_X) and (Y, d_Y) be metric spaces. A map $f : X \to Y$ is called an isometric embedding if for all $x, y \in X$,

$$d_X(x,y) = d_Y(f(x), f(y)).$$

A metric space X is said to be isometrically embedded in a metric space Y if there is an isometric embedding from X to Y. Note that an isometric embedding $f : X \to Y$ is always injective, since if $f(x) = f(y)$, then $d_X(x,y) = d_Y(f(x), f(y)) = 0$. This shows that $x = y$. In Example 2.5.16, X and Y are isometrically embedded in $Z = X \sqcup Y$, which are indeed subspaces of Z. In Example 2.5.16, we assumed that $X \cap Y = \emptyset$. If $X \cap Y \ne \emptyset$, then we may consider the sets $X' = X \times \{0\}$ and $Y' = Y \times \{1\}$. Define d'_X on X' by $d'_X((x,0),(y,0)) = d_X(x,y)$. Then d'_X is a metric on X'. Similarly, we define a metric d'_Y on Y'. Note that $f : X \to X'$ defined by $f(x) = (x,0)$ is an isometric embedding, which is also surjective. Similarly, there is a surjective isometric embedding from Y to Y'. Note that $X' \cap Y' = \emptyset$.

Definition 2.5.18. Let X and Y be bounded metric spaces. Consider the collection

$$\mathcal{A} = \{(Z,f,g) \mid f : X \to Z \text{ and } g : Y \to Z \text{ are isometric embeddings}\}.$$

Then

$$d_{\mathrm{GH}} = \inf\{r \in \mathbb{R} \mid (Z,f,g) \in \mathcal{A} \text{ such that } d_H(f(X), f(Y)) \le r\}$$

is called the Gromov–Hausdorff distance between X and Y.

The collection $\mathcal{A}$ in the above definition is not a set, but the collection to define d_{GH} is a set. Now we will prove that to obtain d_{GH}, we should not look at arbitrary metric spaces Z satisfying the above condition. Let d_X and d_Y be metrics on X and Y, respectively. Let $\mathcal{B}$ denote the set of all metrics d on $X \sqcup Y$ such that $d\,|_X = d_X$ and $d\,|_Y = d_Y$.

Proposition 2.5.19. *Let X and Y be bounded metric spaces. Then*

$$d_{\mathrm{GH}}(X,Y) = \inf\{d_H(X,Y) \mid d \in \mathcal{B}\}.$$

Proof. Let $d_{\mathcal{B}} = \inf\{d_H(X,Y) \mid d \in \mathcal{B}\}$. For every $d \in \mathcal{B}$, we have $(X \sqcup Y, i_X, i_Y) \in \mathcal{A}$, where i_X is the inclusion map. This shows that

$$d_{\mathrm{GH}}(X,Y) \le d_{\mathcal{B}}.$$

Let $\epsilon > 0$. By the definition of the Gromov–Hausdorff distance there is $(Z, f, g) \in \mathcal{A}$ such that

$$d_H(f(X), f(Y)) \le d_{\mathrm{GH}}(X,Y) + \epsilon.$$

Suppose that $f(X) \cap f(Y) = \emptyset$. We may suppose that $Z = f(X) \sqcup g(Y)$. If we identify X with $f(X)$ and Y with $g(Y)$, then we get a metric $d \in \mathcal{B}$ such that

$$d_H(X,Y) \le d_{\mathrm{GH}}(X,Y) + \epsilon.$$

Suppose that $f(X) \cap f(Y) \ne \emptyset$. Consider $Z \times \mathbb{R}$ with the metric $d'((z_1, r_1), (z_2, r_2)) = d(z_1, z_2) + |r_1 - r_2|$. Note that Z is isometrically embedded in $Z \times \mathbb{R}$. Consider $X_1 = f(X) \times \{0\}$ and $Y_1 = g(Y) \times \{\epsilon\}$ in $Z \times \mathbb{R}$. Note that $X_1 \cap Y_1 = \emptyset$. Also, note that

$$d'_H(X_1, Y_1) = d_H(f(X), g(Y)) + \epsilon \le d_{\mathrm{GH}}(X,Y) + 2\epsilon.$$

Identifying X with X_1 and Y with Y_1, we get a metric $\rho \in \mathcal{B}$ such that

$$\rho_H(X,Y) \le d_{\mathrm{GH}}(X,Y) + 2\epsilon.$$

Since $\epsilon > 0$ is arbitrary,

$$d_{\mathcal{B}} \le d_{\mathrm{GH}}(X,Y).$$

Thus $d_{\mathcal{B}} = d_{\mathrm{GH}}(X,Y)$. □

Exercises

2.1.   First of all, complete whatever is left for you as exercises.
2.2.   Let (X, d) be a metric space. Check which one of following is a metric on X:
   (i) $d_1(x,y) = a d(x,y)$ for $a \in \mathbb{R}$,
   (ii) $d_2(x,y) = 2^{d(x,y)}$,
   (iii) $d_3(x,y) = 2^{-d(x,y)}$.

2.3. Let $x = (x_1, \ldots, x_n), y = (y_1, \ldots, y_n), z = (z_1, \ldots, z_n)$, and $w = (w_1, \ldots, w_n)$ be points in the Euclidean space $(\mathbb{R}^n, d)$. Suppose $z_i = \sum_{j=1}^{n} a_{ij} x_j$ and $w_i = \sum_{j=1}^{n} a_{ij} y_j$, where

$$\sum_{i=1}^{n} a_{ik} a_{il} = \begin{cases} 0 & \text{if } k \neq l, \\ 1 & \text{if } k = l. \end{cases}$$

Show that $d(x, y) = d(z, w)$.

2.4. Show that the function $d : \mathbb{C} \times \mathbb{C} \to \mathbb{R}$ defined by

$$d(z, w) = \frac{|z - w|}{\sqrt{(1 + |z|^2)(1 + |w|^2)}}$$

is a metric on $\mathbb{C}$. Also, show that the metric d can be extended to the extended complex plane $\mathbb{C} \cup \{\infty\}$ by defining

$$d(z, \infty) = \frac{1}{1 + |z|^2} \quad \text{and} \quad d(\infty, \infty) = 0.$$

2.5. Consider the metric d_∞ on $\mathbb{R}^2$ and sketch the open ball $B((0, 1), 1)$.

2.6. Show that $d(x, y) = |x - y|^a$ is a metric on $\mathbb{R}$ if $0 < a \leq 1$.

2.7. Let $a_1, \ldots, a_n$ be fixed positive real numbers. Show that

$$d(x, y) = \sum_{i=1}^{n} a_i |x_i - y_i|$$

is a metric on $\mathbb{R}^n$, where $x = (x_1, \ldots, x_n)$ and $y = (y_1, \ldots, y_n)$.

2.8. Show that $\{(x, y, z) \in R^3 \mid x^2 + y^2 \leq z\}$ is a convex set.

2.9. Is arbitrary intersection of convex sets a convex set?

2.10. Show that for $1 \leq p < q \leq \infty$, ℓ_p is contained in ℓ_q.

2.11. Let $x = (x_n) \in \ell_p$ for some $1 \leq p < \infty$. Show that $\|x\|_\infty = \lim_{p \to \infty} \|x\|_p$.

2.12. Let $1 \leq p < \infty$. Define $\| \cdot \|_p$ on $C_{\mathbb{F}}[a, b]$ by

$$\|f\|_p = \left(\int_a^b |f(t)|^p \right)^{\frac{1}{p}}.$$

Show that $\| \cdot \|_p$ is a norm on $C_{\mathbb{F}}[a, b]$.

2.13. Show that $\| \cdot \|_\infty$ is a norm on $C_{\mathbb{F}}[a, b]$, where $\|f\|_\infty = \sup\{|f(t)| \mid t \in [a, b]\}$.

2.14. Let $u > 0$. Define $\|f\| = \min\{\|f\|_\infty, u\|f\|_1\}$ on $C[a, b]$. Find the condition on u such that $\| \cdot \|$ is a norm on $C[a, b]$.

2.15. Let M denote the set of all $m \times n$ real matrices. For every $A = (a_{ij}) \in M$, define

$$\|A\| = \max\left\{ \frac{\|Ax\|}{\|x\|} \,\Big|\, x \in \mathbb{R}^n \setminus \{0\} \right\}.$$

Show that $\|A\|$ defines a norm on M. Also, show that if A is an $n \times n$ matrix, then

$$\|A\| = \max\{y^t A x \mid \|x\| = \|y\| = 1\}.$$

2.16. Compute $\|T\|$, where
  (i) $T : \ell_2 \to \ell_2$ defined by $T((x_n)) = (x_n + x_{n+1})$;
  (ii) $T : (C[0,1], \|\cdot\|_\infty) \to (C[0,1], \|\cdot\|_\infty)$ defined by $T(f(x)) = \int_0^x (x - t)f(t)dt$.

2.17. Which norm $\|\cdot\|_p$ on $C[a,b]$ is induced from an inner product?

2.18. Let $f : [0,\infty) \to [0,\infty)$ be such that f is differentiable on (u,∞) for some $u \geq 0$ and $\lim_{x\to\infty} f'(x) = \infty$. Show that f is not a metric-preserving map.

2.19. If f and g are metric-preserving maps and $k > 0$, then show that $f + g, f \circ g, kf$, and $\max\{f,g\}$ are metric-preserving maps.

2.20. If (f_n) is a sequence of metric-preserving maps and f_n converges to f pointwise with $f(x) > 0$ for all x, then show that f is metric preserving.

2.21. Let f be a metric-preserving map. Show that f is discontinuous at 0 if and only if $f \circ d$ is the discrete metric for every metric d.

2.22. Show that each finite set in a metric space is closed.

2.23. Show that the set A is dense in a metric space X if and only if the complement of A has an empty interior.

2.24. Show that the set $A = \{a + b\sqrt{2} \mid a, b \in \mathbb{Z}\}$ is dense in the real line.

2.25. Show that any subgroup of $(\mathbb{R}, +)$ is either discrete or dense in the real line.

2.26. Let V be a normed space. Let A be a set in V such that $V \setminus A$ is a subspace of V. Show that A is either empty or dense in V.

2.27. Let W be a subspace of a normed space V. Show that $\text{Int } W = \emptyset$ or $W = V$.

2.28. Let $m < n$. Show that the set of all $n \times m$ real matrices of rank m is open in (M_{nm}, d), where M_{nm} is the set of all $n \times m$ real matrices, and

$$d((a_{ij}), (b_{ij})) = \sum_i \sum_j |a_{ij} - b_{ij}|.$$

2.29. Show that the set of all nilpotent matrices is closed in (M_n, d), where M_n is the set of all $n \times n$ real matrices, and d is the metric on M_n as defined in Exercise 2.28.

2.30. Show that ℓ_∞ is not separable.

2.31. Let A and B be nonempty sets in a metric space. Show that

$$d_H(A, B) = d_H(\text{Cl } A, B) = d_H(A, \text{Cl } B).$$

2.32. Let A, B, C, and D be nonempty closed and bounded sets in a metric space such that $C \subseteq A$ and $D \subseteq B$. Show that

$$d_H(A \cup D, B \cup C) \leq d_H(A, B).$$

2.33. Let X and Y be separable metric spaces. Show that

$$d_{\mathrm{GH}}(X, Y) = \inf\{d_h(f(X), g(Y)) \mid f : X \to \ell_\infty, g : Y \to \ell_\infty$$
$$\text{are isometric embeddings}\}.$$

Here the infimum is taken over all the isometric embeddings.

3 Maps between metric spaces

In this chapter, we study continuous maps between metric spaces. We also study homeomorphisms and isometries between metric spaces.

3.1 Continuous maps

Let us recall the ϵ-δ definition of a continuous map $f : \mathbb{R} \to \mathbb{R}$. A map $f : \mathbb{R} \to \mathbb{R}$ is said to be continuous at $a \in \mathbb{R}$ if for each $\epsilon > 0$, there is $\delta > 0$ such that

$$|x - a| < \delta \Rightarrow |f(x) - f(a)| < \epsilon.$$

This definition says that the distance between $f(x)$ and $f(a)$ is closed enough whenever the distance between x and a is closed enough. We adopt this notion in a metric space as follows.

Definition 3.1.1. Let (X, d_X) and (Y, d_Y) be metric spaces. Then a map $f : X \to Y$ is said to be continuous at $a \in X$ if for each $\epsilon > 0$, there is $\delta > 0$ such that

$$d_X(x, a) < \delta \Rightarrow d_Y(f(x), f(a)) < \epsilon.$$

This definition may be reformulated as follows.

A map $f : X \to Y$ is said to be continuous at $a \in X$ if for each open ball $B(f(a), \epsilon)$ around $f(a)$, there is an open ball $B(a, \delta)$ around a such that

$$f(B(a, \delta)) \subseteq B(f(a), \epsilon).$$

We can easily observe that a map $f : X \to Y$ is continuous at $a \in X$ if and only if for each open set V in Y containing $f(a)$, there is an open set U containing a such that $f(U) \subseteq V$.

A map $f : X \to Y$ is called continuous if it is continuous at each point of X. A map that is not continuous is called a discontinuous map.

Example 3.1.2. A constant map between metric spaces is continuous.

Example 3.1.3. The identity map on a metric space X is continuous.

Theorem 3.1.4. *A map $f : X \to Y$ is continuous if and only if the inverse image of an open set in Y is open in X.*

Proof. Let $f : X \to Y$ be continuous, and let U be an open set in Y. If $U = \emptyset$, then $f^{-1}(U) = \emptyset$, and the result follows. Suppose that $U \neq \emptyset$. We will show that each point of $f^{-1}(U)$ is an interior point.

Let $x \in f^{-1}(U)$. Then $f(x) \in U$. Since U is open, there is $\epsilon > 0$ such that

$$B(f(x), \epsilon) \subseteq U.$$

https://doi.org/10.1515/9783111636085-003

Since f is continuous, there is $\delta > 0$ such that

$$f(B(x, \delta)) \subseteq B(f(x), \epsilon).$$

This implies that

$$B(x, \delta) \subseteq f^{-1}(f(B(x, \delta))) \subseteq f^{-1}(B(f(x), \epsilon)) \subseteq f^{-1}(U).$$

This shows that x is an interior point of $f^{-1}(U)$. Since $x \in f^{-1}(U)$ is arbitrary, $f^{-1}(U)$ is open.

Conversely, suppose that $f^{-1}(U)$ is open in X for each open set U in Y. To prove that f is continuous, we will prove that f is continuous at every point of X. Let $x \in X$. Choose an open ball $B(f(x), \epsilon)$ around $f(x)$ in Y. Since an open ball is an open set, $f^{-1}(B(f(x), \epsilon))$ is an open set in X containing x. Then there is a real number $\delta > 0$ such that

$$B(x, \delta) \subseteq f^{-1}(B(f(x), \epsilon)).$$

This implies that

$$f(B(x, \delta)) \subseteq B(f(x), \epsilon).$$

Hence f is continuous at x. Since $x \in X$ is arbitrary, f is continuous. □

Corollary 3.1.5. *A map $f : X \to Y$ is continuous if and only if the inverse image of an open ball in Y is an open set in X.*

Example 3.1.6. Consider the map $f : \mathbb{R} \to \mathbb{R}$ defined by

$$f(x) = \begin{cases} 1 & \text{if } x \geq 0, \\ 0 & \text{if } x < 0. \end{cases}$$

Let $U = (0, 3)$. Since $f^{-1}(U) = [0, \infty)$ is not open in $\mathbb{R}$, f is not continuous.

Note that for each subset A of Y,

$$X \setminus f^{-1}(Y \setminus A) = X \setminus (X \setminus f^{-1}(A)) = f^{-1}(A).$$

Therefore by Theorem 3.1.4 we get the following:

Proposition 3.1.7. *A map $f : X \to Y$ is continuous if and only if the inverse image of a closed set in Y is closed in X.*

Proposition 3.1.8. *A map $f : X \to Y$ is continuous if and only if for each set A in X,*

$$f(\mathrm{Cl}\, A) \subseteq \mathrm{Cl}\, f(A).$$

Proof. Let f be a continuous map. Since $f(A) \subseteq \mathrm{Cl}f(A)$,

$$A \subseteq f^{-1}(f(A)) \subseteq f^{-1}(\mathrm{Cl}f(A)).$$

Since f is continuous, $f^{-1}(\mathrm{Cl}f(A))$ is a closed set in X. Therefore

$$\mathrm{Cl}A \subseteq f^{-1}(\mathrm{Cl}f(A)).$$

Hence

$$f(\mathrm{Cl}A) \subseteq f(f^{-1}(\mathrm{Cl}f(A))) \subseteq \mathrm{Cl}f(A).$$

Conversely, suppose that for each set A in X,

$$f(\mathrm{Cl}A) \subseteq \mathrm{Cl}f(A).$$

Let F be a closed set in Y. We will show that $f^{-1}(F)$ is a closed set in X.
    By the assumption,

$$f(\mathrm{Cl}f^{-1}(F)) \subseteq \mathrm{Cl}f(f^{-1}(F)).$$

Since $f(f^{-1}(F)) \subseteq F$,

$$f(\mathrm{Cl}f^{-1}(F)) \subseteq \mathrm{Cl}F.$$

Since F is closed,

$$f(\mathrm{Cl}f^{-1}(F)) \subseteq F.$$

Therefore

$$\mathrm{Cl}f^{-1}(F) \subseteq f^{-1}(f(\mathrm{Cl}f^{-1}(F))) \subseteq f^{-1}(F).$$

This shows that

$$\mathrm{Cl}f^{-1}(F) = f^{-1}(F).$$

Hence $f^{-1}(F)$ is a closed set in X. □

Proposition 3.1.9. *A map $f : X \to Y$ is continuous if and only if for each set B in Y,*

$$f^{-1}(\mathrm{Int}\,B) \subseteq \mathrm{Int}\,f^{-1}(B).$$

Proof. Suppose f is continuous. Since $\mathrm{Int}\,B$ is an open set in Y, $f^{-1}(\mathrm{Int}\,B)$ is open in X. Since $\mathrm{Int}\,B \subseteq B$, $f^{-1}(\mathrm{Int}\,B) \subseteq f^{-1}(B)$. This implies that

$$f^{-1}(\operatorname{Int} B) \subseteq \operatorname{Int} f^{-1}(B).$$

Conversely, suppose that for all sets B in Y,

$$f^{-1}(\operatorname{Int} B) \subseteq \operatorname{Int} f^{-1}(B).$$

Let U be an open set in Y. Then $\operatorname{Int} U = U$. By the assumption we have

$$f^{-1}(U) = f^{-1}(\operatorname{Int} U) \subseteq \operatorname{Int} f^{-1}(U) \subseteq f^{-1}(U).$$

Therefore $\operatorname{Int} f^{-1}(U) = f^{-1}(U)$. This shows that $f^{-1}(U)$ is open in X. Hence f is continuous. □

Corollary 3.1.10. *A map $f : X \to Y$ is continuous if and only if for each set B in Y,*

$$\operatorname{Cl} f^{-1}(B) \subseteq f^{-1}(\operatorname{Cl} B).$$

Proof. Let $C = Y \setminus B$. By Proposition 3.1.9 we have

$$
\begin{aligned}
f \text{ is continuous} &\Leftrightarrow f^{-1}(\operatorname{Int} C) \subseteq \operatorname{Int} f^{-1}(C)\\
&\Leftrightarrow f^{-1}(\operatorname{Int}(Y \setminus B)) \subseteq \operatorname{Int} f^{-1}(Y \setminus B)\\
&\Leftrightarrow f^{-1}(Y \setminus \operatorname{Cl} B) \subseteq \operatorname{Int}(X \setminus f^{-1}(B))\\
&\Leftrightarrow X \setminus f^{-1}(\operatorname{Cl} B) \subseteq \operatorname{Int}(X \setminus f^{-1}(B))\\
&\Leftrightarrow X \setminus f^{-1}(\operatorname{Cl} B) \subseteq X \setminus \operatorname{Cl} f^{-1}(B)\\
&\Leftrightarrow \operatorname{Cl} f^{-1}(B) \subseteq f^{-1}(\operatorname{Cl} B).
\end{aligned}
$$
□

Corollary 3.1.11. *A map $f : X \to Y$ is continuous if and only if for each set B in Y,*

$$\operatorname{Bd} f^{-1}(B) \subseteq f^{-1}(\operatorname{Bd} B).$$

Proof. Suppose that f is continuous. Then by Corollary 3.1.10

$$
\begin{aligned}
\operatorname{Bd} f^{-1}(B) &= \operatorname{Cl} f^{-1}(B) \cap \operatorname{Cl}(X \setminus f^{-1}(B))\\
&= \operatorname{Cl} f^{-1}(B) \cap \operatorname{Cl} f^{-1}(Y \setminus B)\\
&\subseteq f^{-1}(\operatorname{Cl} B) \cap f^{-1}(\operatorname{Cl}(Y \setminus B))\\
&= f^{-1}(\operatorname{Cl} B \cap \operatorname{Cl}(Y \setminus B))\\
&= f^{-1}(\operatorname{Bd} B).
\end{aligned}
$$

Conversely, suppose that for each set B in Y,

$$\operatorname{Bd} f^{-1}(B) \subseteq f^{-1}(\operatorname{Bd} B).$$

Let U be an open set in Y. Then $Y \setminus U$ is closed in Y. By Proposition 2.4.55 we have

$$\mathrm{Bd}(Y \setminus U) \subseteq Y \setminus U.$$

Therefore

$$f^{-1}(\mathrm{Bd}(Y \setminus U)) \subseteq f^{-1}(Y \setminus U).$$

By the assumption,

$$\mathrm{Bd}\, f^{-1}(Y \setminus U) \subseteq f^{-1}(\mathrm{Bd}(Y \setminus U)).$$

Therefore

$$\mathrm{Bd}\, f^{-1}(Y \setminus U) \subseteq f^{-1}(Y \setminus U).$$

Therefore $f^{-1}(Y \setminus U)$ is closed by Proposition 2.4.55. Hence $f^{-1}(U)$ is open in X. This shows that f is continuous. $\qquad\square$

By Proposition 3.1.9 a map $f : X \to Y$ is continuous if and only if for all $a \in X$ and for all sets B in Y,

$$f(a) \in \mathrm{Int}\, B \Rightarrow a \in \mathrm{Int}\, f^{-1}(B).$$

Example 3.1.12. Consider the map $f : \mathbb{R} \to \mathbb{R}$ defined by

$$f(x) = \begin{cases} \frac{x^2-1}{x-1} & \text{if } x \neq y, \\ 3 & \text{if } x = y. \end{cases}$$

Let $B = (2, 4)$. Note that

$$f(1) = 3 \in B.$$

Let $x \in f^{-1}(B)$ be such that $x \neq 1$. Then

$$f(x) = x + 1 \in B.$$

This implies that $x \in (1, 3)$. This shows that

$$f^{-1}(B) \subseteq \{1\} \cup (1, 3) = [1, 3).$$

Since $1 \notin \mathrm{Int}\, f^{-1}(B)$, f is not continuous.

We have seen that the constant map $f : X \to Y$ defined by $f(x) = c$ for all $x \in X$ is continuous, where $c \in Y$ is fixed element. Moreover, for any open set U in Y, we have

$$f^{-1}(U) = \begin{cases} \emptyset & \text{if } c \notin U, \\ X & \text{if } c \in U. \end{cases}$$

Example 3.1.13. Every map f from the discrete metric space X to any metric space Y is continuous.

Example 3.1.14. Let A be a subspace of a metric space X. Then the inclusion map $i : A \to X$ is continuous, since if U is open in X, then $i^{-1}(U) = U \cap A$ is open in A.

Example 3.1.15. The identity map $I : X \to X$ need not be continuous if the domain and codomain are with different metrics. For example, consider the set $\mathbb{R}$ of real numbers with the discrete metric d_1 and the usual metric d_2. Then the identity map $I : (X, d_2) \to (X, d_1)$ is not continuous. Note that $\{0\}$ is open in (X, d_1) but $\{0\} = I^{-1}(\{0\})$ is not open in (X, d_2).

This example shows that the inverse map of a bijective map between metric spaces need not be continuous.

Proposition 3.1.16. *Let d_1 and d_2 be two metrics on a set X. Then the identity map $I : (X, d_1) \to (X, d_2)$ is continuous if and only if for each open ball $B_{d_2}(x, \epsilon)$ in (X, d_2), there is an open ball $B_{d_1}(x, \delta)$ in (X, d_1) such that $B_{d_1}(x, \delta) \subseteq B_{d_2}(x, \epsilon)$.*

Proof. Suppose that the identity map $I : (X, d_1) \to (X, d_2)$ is continuous. Consider an open ball $B_{d_2}(x, \epsilon)$ in (X, d_2). Then

$$B_{d_2}(x, \epsilon) = I^{-1}(B_{d_2}(x, \epsilon))$$

is open in (X, d_1). Then there is a real number $\delta > 0$ such that

$$B_{d_1}(x, \delta) \subseteq B_{d_2}(x, \epsilon).$$

For the converse, let U be an open set in (X, d_2). If $U = \emptyset$, then $I^{-1}(U)$ is open in (X, d_1). Suppose that $U \neq \emptyset$. Let $x \in U$. Then there is a real number $\epsilon > 0$ such that

$$B(x, \epsilon) \subseteq U.$$

By the assumption there is an open ball $B_{d_1}(x, \delta)$ in (X, d_1) such that

$$B_{d_1}(x, \delta) \subseteq B_{d_2}(x, \epsilon) \subseteq U = I^{-1}(U).$$

This shows that I is continuous. $\qquad\qquad\square$

Proposition 3.1.17. *Let X, Y, and Z be metric spaces. Let $f : X \to Y$ and $g : Y \to Z$ be continuous at $a \in X$ and $f(a) \in Y$, respectively. Then $g \circ f : X \to Z$ is continuous at $a \in X$.*

Proof. Consider an open ball $B((g \circ f)(a), \epsilon)$ around $g \circ f(a)$ in Z. Since g is continuous at $f(a)$, there is an open ball $B(f(a), \delta')$ around $f(a)$ such that

$$g(B(f(a), \delta')) \subseteq B((g \circ f)(a), \epsilon). \tag{3.1}$$

Since f is continuous at a, there is an open ball $B(x, \delta)$ around a such that

$$f(B(a, \delta)) \subseteq B(f(a), \delta'). \tag{3.2}$$

By equations (3.1) and (3.2) we have

$$(g \circ f)(B(a, \delta)) \subseteq B(g \circ f(a), \epsilon).$$

Hence $g \circ f$ is continuous at $a \in X$. $\qquad\square$

Corollary 3.1.18. *The composition of continuous maps is continuous.*

Example 3.1.19. If $f : X \to Y$ is a continuous map and A is a subspace of X, then the restriction map $f \mid_A : A \to Y$ is continuous, since $f \mid_A = f \circ i$, where $i : A \to X$ is the inclusion map.

Example 3.1.20. The restriction of a discontinuous map may be continuous. For example, consider the map $f : \mathbb{R} \to \mathbb{R}$ defined by

$$f(x) = \begin{cases} 0 & \text{if } x \in \mathbb{Q}, \\ 1 & \text{if } x \in \mathbb{R} \setminus \mathbb{Q}. \end{cases}$$

Note that $\{0\}$ is closed in $\mathbb{R}$ and $f^{-1}(\{0\}) = \mathbb{Q}$. Since $\mathbb{Q}$ is not closed in $\mathbb{R}$, f is not continuous. We can observe that $f \mid_{\mathbb{Q}} : \mathbb{Q} \to \mathbb{R}$ is continuous, since $f \mid_{\mathbb{Q}}$ is constant map.

Proposition 3.1.21 (Pasting lemma). *Let X and Y be metric spaces, and let A and B be closed sets of X such that $A \cup B = X$. Let $f : A \to Y$ and $g : B \to Y$ be continuous maps such that $f(x) = g(x)$ for all $x \in A \cap B$. Then the map $h : X \to Y$ defined by*

$$h(x) = \begin{cases} f(x), & x \in A, \\ g(x), & x \in B, \end{cases}$$

is continuous.

Proof. Let F be a closed set in Y. Then $f^{-1}(F)$ and $g^{-1}(F)$ are closed in A and B, respectively. Since A and B are closed sets in X, $f^{-1}(F)$ and $g^{-1}(F)$ are closed in X. Note that

$$h^{-1}(F) = f^{-1}(F) \cup g^{-1}(F).$$

This shows that $h^{-1}(F)$ is closed in X. Hence h is continuous. $\qquad\square$

Example 3.1.22. Define $h : \mathbb{R} \to \mathbb{R}$ by

$$h(x) = |x| = \begin{cases} -x & \text{if } x \leq 0, \\ x & \text{if } x \geq 0. \end{cases}$$

Let $A = (-\infty, 0]$ and $B = [0, \infty)$. Note that the maps $f : A \to \mathbb{R}$ and $g : B \to \mathbb{R}$ defined by $f(x) = -x$ and $g(x) = x$ are continuous. By the pasting lemma, h is continuous.

We will now focus on the real-valued continuous maps.

Proposition 3.1.23. *Let X be a metric space. Then a map $f : X \to \mathbb{R}$ is continuous if and only if $f^{-1}(a, \infty)$ and $f^{-1}(-\infty, b)$ are open sets of X for all $a, b \in \mathbb{R}$.*

Proof. Note that

$$(a, b) = (-\infty, b) \cap (a, \infty).$$

By Corollary 3.1.5, f is continuous. $\qquad\square$

Example 3.1.24. Let $k \in \mathbb{R}$. Then the map $f : \mathbb{R} \to \mathbb{R}$ defined by $f(x) = kx$ is continuous, since if $k = 0$, then f is constant. If $k \neq 0$, then

$$f^{-1}(a, \infty) = \{x \in \mathbb{R} \mid f(x) > a\}$$
$$= \begin{cases} (\frac{a}{k}, \infty) & \text{if } k > 0, \\ (-\infty, \frac{a}{k}) & \text{if } k < 0, \end{cases}$$

and

$$f^{-1}(-\infty, b) = \{x \in \mathbb{R} \mid f(x) < b\}$$
$$= \begin{cases} (-\infty, \frac{b}{k}) & \text{if } k > 0, \\ (\frac{b}{k}, \infty) & \text{if } k < 0. \end{cases}$$

Proposition 3.1.25. *Let (X, d) be a metric space. Let f and g be continuous maps from X to $\mathbb{R}$. Then*
(i)   *$f + g : X \to \mathbb{R}$ defined by $(f + g)(x) = f(x) + g(x)$ is continuous,*
(ii) *$fg : X \to \mathbb{R}$ defined by $(fg)(x) = f(x)g(x)$ is continuous,*
(iii) *if $g(x) \neq 0$ for all $x \in X$, then $\frac{f}{g} : X \to \mathbb{R}$ defined by $(\frac{f}{g})(x) = \frac{f(x)}{g(x)}$ is continuous.*

Proof. We will prove (i). The rest is left as an exercise.

Let $\epsilon > 0$ and $a \in X$. Since f is continuous, there is $\delta_1 > 0$ such that

$$d(x, a) < \delta_1 \Rightarrow |f(x) - f(a)| < \frac{\epsilon}{2}.$$

Since g is continuous, there is $\delta_2 > 0$ such that

$$d(x, a) < \delta_2 \Rightarrow |g(x) - g(a)| < \frac{\epsilon}{2}.$$

For $\delta = \min\{\delta_1, \delta_2\}$, we have

$$
\begin{aligned}
|(f + g)(x) - (f + g)(a)| &= |(|f(x) - f(a)|) + (|g(x) - g(a)|)| \\
&\leq \frac{\epsilon}{2} + \frac{\epsilon}{2} \\
&= \epsilon
\end{aligned}
$$

whenever $d(x, a) < \delta$. This shows that $f + g$ is continuous. □

Example 3.1.26. By Proposition 3.1.25 we can observe that the map $f : \mathbb{R} \to \mathbb{R}$ defined by

$$f(x) = a_0 + a_1 x + \cdots + a_n x^n$$

is continuous, where $n \in \mathbb{N}$ and $a_i \in \mathbb{R}$.

Let us consider the map $f : \mathbb{R} \to \mathbb{R}$ defined by $f(x) = x^2$. Note that f is continuous. Take $\epsilon = 1$. Since f is continuous at $a = 0$, there is $\delta > 0$ such that

$$|x| < \delta \Rightarrow |x^2| < 1. \tag{3.3}$$

Notes that equation (3.3) is satisfied if we take $\delta = 1$. Now let us deal with the continuity of f at $a = 2$. Let $x = 2 + \frac{1}{3}$. Then $|x - 2| = \frac{1}{3} < 1$, but

$$|f(x) - f(2)| = |x^2 - 2^2| > 1.$$

This shows that the value of δ that works for $a = 0$ does not work for $a = 2$. We are interested in the case of a continuous map when the value of δ uniformly works for all points of a metric space for a given ϵ.

Definition 3.1.27. Let (X, d_X) and (Y, d_Y) be metric spaces. A map $f : X \to Y$ is called a uniformly continuous map if for each $\epsilon > 0$, there exists $\delta > 0$ such that for all $x, y \in X$, we have

$$d_X(x, y) < \delta \Rightarrow d_Y(f(x), f(y)).$$

From the definition note that the continuity is a local concept, which is defined at a particular point, whereas the uniform continuity is not a local concept. We can easily observe that a uniformly continuous map is continuous. The map $f : \mathbb{R} \to \mathbb{R}$ defined by $f(x) = x^2$ is continuous but not uniformly continuous. Also, the constant and identity maps are uniformly continuous.

Proposition 3.1.28. *Let (X, d_X) and (Y, d_Y) be metric spaces. Let $f : X \to Y$ be a map such that for all $x, y \in X$,*

$$d_Y(f(x), f(y)) \le k d_X(x, y)$$

for some positive real number k. Then f is uniformly continuous.

Proof. Let $\epsilon > 0$. Then

$$d_X(x, y) < \frac{\epsilon}{k} \Rightarrow d_Y(f(x), f(y)) < \epsilon.$$

This shows that f is uniformly continuous. □

Example 3.1.29. Consider the map $f : \mathbb{R} \to \mathbb{R}$ defined by

$$f(x) = \frac{x^2}{x^2 + 1}.$$

We can observe that $|x| < x^2 + 1$ for all $x \in \mathbb{R}$ by considering the cases $|x| < 1$ and $|x| \ge 1$. Now

$$
\begin{aligned}
|f(x) - f(y)| &= \left| \frac{x^2}{x^2 + 1} - \frac{y^2}{y^2 + 1} \right| \\
&= \frac{|x - y||x + y|}{(x^2 + 1)(y^2 + 1)} \\
&\le \frac{|x - y|(|x| + |y|)}{(x^2 + 1)(y^2 + 1)} \\
&\le \frac{|x - y|((x^2 + 1) + (y^2 + 1))}{(x^2 + 1)(y^2 + 1)} \\
&= |x - y| \left(\frac{1}{x^2 + 1} + \frac{1}{y^2 + 1} \right) \\
&\le 2|x - y|.
\end{aligned}
$$

Therefore by Proposition 3.1.28 f is uniformly continuous.

Example 3.1.30. Let (X, d) be a metric space, and let A be a fixed nonempty set in X. Consider the map $f : X \to \mathbb{R}$ defined by $f(x) = d(x, A)$. By Proposition 2.5.3, for $x, y \in X$, we have

$$d(x, A) - d(y, A) \le d(x, y).$$

Interchanging the role of x and y, we get

$$d(y, A) - d(x, A) \le d(x, y).$$

This shows that

$$|f(x) - f(y)| = |d(x,A) - d(y,A)| \le d(x,y).$$

By Proposition 3.1.28 f is uniformly continuous.

Proposition 3.1.31 (Urysohn's lemma). *Let F_1 and F_2 be disjoint closed subsets of a metric space X. Then there is a continuous map $f : X \to [-1,1]$ such that*

$$f^{-1}(\{-1\}) = F_1 \quad and \quad f^{-1}(\{1\}) = F_2.$$

Proof. Define the map $f : X \to [-1,1]$ by

$$f(x) = \frac{d(x,F_1) - d(x,F_2)}{d(x,F_1) + d(x,F_2)}.$$

First, observe that $d(x,F_1) + d(x,F_2) \ne 0$ for all $x \in X$. On the contrary, suppose that there is $x \in X$ such that

$$d(x,F_1) + d(x,F_2) = 0.$$

Then $d(x,F_1) = 0$ and $d(x,F_2) = 0$. This implies that $x \in \mathrm{Cl}\,F_1 \cap \mathrm{Cl}\,F_2 = F_1 \cap F_2 = \emptyset$, a contradiction. By Example 3.1.30 and Proposition 3.1.25 f is continuous.
     Note that

$$f(x) = -1 \Leftrightarrow d(x,F_1) = 0 \Leftrightarrow x \in \mathrm{Cl}\,F_1 = F_1$$

and

$$f(x) = 1 \Leftrightarrow d(x,F_2) = 0 \Leftrightarrow x \in \mathrm{Cl}\,F_2 = F_2. \qquad \square$$

Recall that for normed spaces V and W over a field $\mathbb{F}$, $B(V,W)$ denotes the set of all linear transformations $T : V \to W$ such that there exists $k > 0$ such that $\|T(x)\| \le k\|x\|$ for all $x \in V$.

Proposition 3.1.32. *Let V and W be normed spaces, and let $T : V \to W$ be a linear transformation. Then the following statements are equivalent:*
(i)   *T is uniformly continuous;*
(ii)   *T is continuous;*
(iii) *T is continuous at 0;*
(iv)   *$T \in B(V,W)$.*

Proof. The implications (i) $\Rightarrow$ (ii) and (ii) $\Rightarrow$ (iii) are obvious. We will now prove (iii) $\Rightarrow$ (iv).

Assume (iii). For $\epsilon = 1$, there exists $\delta > 0$ such that for all $x \in V$,

$$\|x\| < \delta \Rightarrow \|T(x)\| = \|T(x) - T(0)\| < 1.$$

Suppose that $x \neq 0$. Let $y = \frac{x}{\|x\|}$. Note that

$$\left\|\frac{\delta y}{3}\right\| = \frac{\delta}{3} < \delta.$$

This implies that

$$\frac{\delta}{3}\|T(y)\| = \left\|T\left(\frac{\delta y}{3}\right)\right\| < 1.$$

This shows that

$$\|T(x)\| < \frac{3}{\delta}\|x\|.$$

For $x = 0$, the condition $\|T(x)\| \leq \frac{3}{\delta}\|x\|$ trivially holds. Therefore $T \in B(V, W)$.

Assume (iv). Then there is a positive real number k such that $\|T(x)\| \leq k\|x\|$ for all $x \in V$. Now for all $x, y \in V$, we have

$$\|T(x - y)\| = \|T(x) - T(y)\| \leq k\|x - y\|.$$

By Proposition 3.1.28 T is uniformly continuous. □

Let V and W be two normed spaces, and let $T : V \to W$ be a linear transformation. Then the set

$$\mathrm{Ker}\,T = \{x \in V \mid T(X) = 0\}$$

is called the kernel of T. We can check that $\mathrm{Ker}\,T$ is a subspace of V. Note that $\mathrm{Ker}\,T = T^{-1}(\{0\})$. This shows that if the linear transformation T is continuous, then $\mathrm{Ker}\,T$ is a closed subspace of V. Let us observe the converse of this statement in the following case.

Proposition 3.1.33. *Let V be a normed space, and let $f : V \to \mathbb{F}$ be a linear functional. If $\mathrm{Ker}\,f$ is closed, then f is continuous.*

Proof. If $f(x) = 0$ for all $x \in V$, then f is continuous. Suppose that f is a nonzero map. Let $\epsilon > 0$. Let

$$U_\epsilon = \{x \in V \mid |f(x)| < \epsilon\}.$$

Let $w \in V$ be such that $f(w) \neq 0$. Let $u = \frac{w}{f(w)}$. Then $f(u) = 1$. This implies that $u \notin \operatorname{Ker} f$. Therefore $\epsilon u \notin \operatorname{Ker} f$, that is, $\epsilon u \in V \setminus \operatorname{Ker} f$. Since $\operatorname{Ker} f$ is closed, there is an open ball $B(\epsilon u, r)$ contained in $V \setminus \operatorname{Ker} f$.

Let $x \in B(0, r)$. We claim that $x \in U_\epsilon$. On the contrary, suppose that $x \notin U_\epsilon$. Then $|f(x)| \geq \epsilon$. Let $y = \frac{-\epsilon x}{f(x)}$. Then

$$\|y\| = \frac{\epsilon}{|f(x)|} \|x\| \leq \|x\| < r.$$

This implies that $y \in B(0, r)$. Therefore $y + \epsilon u \in B(\epsilon u, r)$. Note that

$$f(y + \epsilon u) = f(y) + \epsilon f(u) = 0.$$

This is a contradiction as $y + \epsilon u \in V \setminus \operatorname{Ker} f$. Hence $x \in U_\epsilon$. Since $x \in B(0, r)$, $B(0, r) \subseteq U_\epsilon$. Equivalently, we have

$$\|x - 0\| < r \implies |f(x) - f(0)| = |f(x)| < \epsilon.$$

Therefore f is continuous at 0. Since f is linear, f is continuous. $\qquad\square$

Theorem 3.1.34. *Let V be a normed space. Then every linear functional $f : V \to \mathbb{F}$ is continuous if and only if every subspace of V is closed.*

Proof. Suppose every subspace of V is closed. Let $f : V \to \mathbb{F}$ be a linear functional. Then $\operatorname{Ker} f$ is closed. By Proposition 3.1.33 f is continuous.

Conversely, suppose that every linear functional $f : V \to \mathbb{F}$ is continuous. Let W be a subspace of V. If $W = V$, then it is closed. Suppose that W is a proper subspace of V. Consider a basis $\{v_\alpha \mid \alpha \in \mathcal{A}\}$ of W. Extend this to a basis

$$\{v_\alpha \mid \alpha \in \mathcal{A}\} \cup \{v_\beta \mid \alpha \in \mathcal{B}\}$$

of V, where $\mathcal{A} \cap \mathcal{B} = \emptyset$. For each $i \in \mathcal{A} \cup \mathcal{B}$, define the linear functional $f_i : V \to \mathbb{F}$ by $f_i(v_j) = \delta_{ij}$, where δ_{ij} is the Kronecker delta function. By assumption each f_i is continuous for each $i \in \mathcal{A} \cup \mathcal{B}$. We claim that

$$W = \bigcap_{i \in \mathcal{B}} \operatorname{Ker} f_i.$$

Clearly, $W \subseteq \operatorname{Ker} f_i$ for all $i \in \mathcal{B}$. Then

$$W \subseteq \bigcap_{i \in \mathcal{B}} \operatorname{Ker} f_i.$$

Conversely, suppose that $x \in \bigcap_{i \in \mathcal{B}} \operatorname{Ker} f_i$. Express $x = \sum_{i \in \mathcal{A} \cup \mathcal{B}} x_i v_i$, where the expression is a finite sum. Since $f_i(x) = 0$ for each $i \in \mathcal{B}$, we get $x = \sum_{i \in \mathcal{A}} x_i v_i$. This implies that $x \in W$. Therefore

$$\bigcap_{i \in B} \operatorname{Ker} f_i \subseteq W.$$

Hence

$$W = \bigcap_{i \in B} \operatorname{Ker} f_i.$$

Since each $\operatorname{Ker} f_i$ is closed, W is closed. □

Definition 3.1.35. Let (X, d_X) and (Y, d_Y) be metric spaces. Let A be a nonempty set in X, and let $a \in X$ be a limit point of A. Then $l \in Y$ is called a limit of a map $f : A \to Y$ as x tends to a if for each $\epsilon > 0$, there exists $\delta > 0$ such that for all $x \in A$,

$$0 < d_X(x, a) < \delta \Rightarrow d_Y(f(x), l) < \epsilon.$$

This definition may be reformulated as follows.

A point $l \in Y$ is called a limit of a map $f : A \to Y$ as x tends to a limit point a of A if for each $\epsilon > 0$, there exists $\delta > 0$ such that

$$f(B(a, \delta) \setminus \{a\}) \subseteq B(l, \epsilon).$$

If $l \in Y$ is a limit of a map $f : A \to Y$ as x tends to a, then we represent this as

$$\lim_{x \to a} f(x) = l$$

or

$$f(x) \to l \quad \text{as } x \to a.$$

Note 3.1.36. Note that a limit of a map $f : A \to Y$ is defined at a limit point of A, which may be outside A.

Example 3.1.37. Let $A = \{(x, y) \mid x^2 + y^2 < 1\} \subseteq \mathbb{R}^2$, and let $f : A \to \mathbb{R}^3$ be defined as

$$f(x, y) = (x^2, 2y, x + y).$$

Then we can observe that

$$\lim_{(x,y) \to (1,0)} f(x, y) = (1, 0, 1).$$

Example 3.1.38. Let $f : (0, 1) \to \mathbb{R}$ be defined by $f(x) = \frac{1}{x}$. Then we can observe that the limit of f does not exist as x tends to 0.

Proposition 3.1.39. *Let X and Y be metric spaces, and let $A \subseteq X$. If a limit of a map $f : A \to Y$ exists, then it is unique.*

Proof. Let a be a limit point of A. On the contrary, suppose that l_1 and l_2 are two different limits of f as x tends to a. This implies that $\epsilon = d(l_1, l_2) > 0$. Then there are real numbers $\delta_1 > 0$ and $\delta_2 > 0$ such that

$$0 < d(x, a) < \delta_1 \Rightarrow d(f(x), l_1) < \frac{\epsilon}{2}$$

and

$$0 < d(x, a) < \delta_2 \Rightarrow d(f(x), l_2) < \frac{\epsilon}{2}.$$

Let $\delta = \min\{\delta_1, \delta_2\}$. Then for $x \in A$ such that $0 < d(x, a) < \delta$, we have

$$\begin{aligned}
\epsilon &= d(l_1, l_2) \\
&\leq d(f(x), l_1) + d(f(x), l_2) \\
&< \frac{\epsilon}{2} + \frac{\epsilon}{2} \\
&= \epsilon,
\end{aligned}$$

a contradiction. $\qquad\square$

Note that we have defined the notion of a limit at a limit point. A map $f : A \to Y$ may be continuous at a point $a \in A$ without having a limit at that point if a is not a limit point of A. Let us see what happens if we had defined the notion of a limit at any point of A.

Let d_1 and d_2 denote the discrete and usual metrics on $\mathbb{R}$, respectively. Define $f : (\mathbb{R}, d_1) \to (\mathbb{R}, d_2)$ by $f(x) = x^2$. Since the domain of f is equipped with the discrete metric, f is continuous map. Note that $D(\mathbb{R}) = \emptyset$ in $(\mathbb{R}, d_1)$. Let us try to find the limit of the map f, say at 0. Note that the open ball $B(0, \frac{1}{2}) = \{0\}$ in $(\mathbb{R}, d_1)$. Let l be any real number, and let $\epsilon > 0$. Then

$$\emptyset = f\left(B\left(0, \frac{1}{2}\right) \setminus \{0\}\right) \subseteq B(l, \epsilon) = (l - \epsilon, l + \epsilon).$$

This shows that all the real numbers are the limits of the map f. This is a contradiction on the uniqueness of the limit if it exists. If $a \in A$ is a limit point and f is continuous at a, then

$$\lim_{x \to a} f(x) = f(a).$$

3.2 Homeomorphisms

In the previous section, we saw that a map $f : X \to Y$ is continuous if and only if the inverse image of each open (closed) set in Y is open (closed) in X. We are now interested in a map that sends an open (closed) set to an open (closed) set.

Definition 3.2.1. Let X and Y be metric spaces. Then $f : X \to Y$ is called an open map if $f(U)$ is open in Y for every open set U in X.

Definition 3.2.2. Let X and Y be metric spaces. Then $f : X \to Y$ is called a closed map if $f(F)$ is closed in Y for every closed set F in X.

Example 3.2.3. Consider the inclusion map $i : [0,1] \to \mathbb{R}$. Note that $[0, \frac{1}{2})$ is open in $[0,1]$ but is not open in $\mathbb{R}$. This shows that i is not an open map. Note that the inclusion map $i : [0,1] \to \mathbb{R}$ is a closed map.

Example 3.2.4. The inclusion map $i : (0,1) \to \mathbb{R}$ is not a closed map but is an open map.

Example 3.2.5. The identity map $I : (X, d_1) \to (X, d)$, where d is the discrete metric, is an open and closed map.

Observe that the composition of open (closed) maps is an open (a closed) map.

Proposition 3.2.6. *Let X and Y be metric spaces. If $f : X \to Y$ is a map such that $f(B)$ is an open set in Y for each open ball B in X, then f is an open map.*

Proof. Let U be an open set in X. Then U is a union of open balls around each point of U that are contained in U. The proof follows from the fact that

$$f\left(\bigcup_{\alpha \in I} A_\alpha\right) = \bigcup_{\alpha \in I} f(A_\alpha),$$

where $\{A_\alpha \mid \alpha \in I\}$ is a family of sets in X. □

Proposition 3.2.7. *Let X and Y be metric spaces. Then $f : X \to Y$ is an open map if and only if for each set A in X,*

$$f(\operatorname{Int} A) \subseteq \operatorname{Int} f(A).$$

Proof. Let $f : X \to Y$ be an open map, and let $A \subseteq X$. Then $f(\operatorname{Int} A)$ is an open set in Y. Since $\operatorname{Int} A \subseteq A$, $f(\operatorname{Int} A) \subseteq f(A)$. This implies that

$$f(\operatorname{Int} A) \subseteq \operatorname{Int} f(A).$$

Conversely, suppose that for each set A in X,

$$f(\operatorname{Int} A) \subseteq \operatorname{Int} f(A).$$

Let U be an open set in X. Then

$$f(U) = f(\operatorname{Int} U) \subseteq \operatorname{Int} f(U) \subseteq f(U).$$

This shows that $f(U)$ is open in Y. Hence f is an open map. □

Using a similar argument, we can prove the following statement. The proof is left as an exercise.

Proposition 3.2.8. *Let X and Y be metric spaces. Then $f : X \to Y$ is a closed map if and only if for each set A in X,*

$$\operatorname{Cl} A \subseteq f(\operatorname{Cl} A).$$

Proposition 3.2.9. *Let f be a closed map from a metric space X to a metric space Y. Let $A \subseteq Y$, and let U be an open set of X such that $f^{-1}(A) \subseteq U$. Then there is an open set V of Y such that*

$$A \subseteq V \quad and \quad f^{-1}(V) \subseteq U.$$

Proof. Let $V = Y \setminus f(X \setminus U)$. Since f is a closed map, $f(X \setminus U)$ is closed in Y. Therefore V is open in Y.

Let $y \in A$. Then

$$f^{-1}(\{y\}) \subseteq f^{-1}(A) \subseteq U.$$

Therefore $y \in f(U)$. This shows that

$$y \notin f(X \setminus U).$$

In other words,

$$y \in Y \setminus f(X \setminus U) = V.$$

Therefore $A \subseteq V$. We can easily observe that

$$f^{-1}(V) \subseteq U. \qquad \square$$

Definition 3.2.10. Let X and Y be metric spaces. Then a bijective map $f : X \to Y$ is called a homeomorphism if both f and $f^{-1} : Y \to X$ are continuous.

The metric spaces X and Y are called homeomorphic if there is a homeomorphism from X to Y. Note the following:

- The identity map between the same metric spaces is a homeomorphism.
- If $f : X \to Y$ is a homeomorphism, then $f^{-1} : Y \to X$ is a homeomorphism.
- The composition of homeomorphisms is a homeomorphism.

Therefore the relation of being homeomorphic is an equivalence relation in a set of metric spaces.

Proposition 3.2.11. *The metric spaces X and Y are homeomorphic if and only if there exist continuous maps*

$$f : X \to Y \quad and \quad g : Y \to X$$

such that

$$g \circ f = I_X \quad and \quad f \circ g = I_Y.$$

Proof. Left as an exercise. □

Proposition 3.2.12. *Let X and Y be metric spaces, and let $f : X \to Y$ be a bijective map. Then f is a homeomorphism if and only if f is a continuous and open map.*

Proof. Suppose that f is a homeomorphism. Let U be an open set in X. Since $f^{-1} : Y \to X$ is continuous, $f(U) = (f^{-1})^{-1}(U)$ is open in Y. Therefore f is open.

Conversely, suppose that f is continuous and open. Let U be an open set in X. Since $(f^{-1})^{-1}(U) = f(U)$ and f is open, $f^{-1} : Y \to X$ is continuous. Therefore f is homeomorphism. □

Since for a bijective map $f : X \to Y$, the notion of an open map is the same as that of a closed map, we have the following:

Proposition 3.2.13. *Let X and Y be metric spaces, and let $f : X \to Y$ be a bijective map. Then f is a homeomorphism if and only if f is a continuous and closed map.*

Example 3.2.14. Let $a, b, c, d \in \mathbb{R}$ be such that $a < b$ and $c < d$. Suppose the open intervals (a, b) and (c, d) are with the subspace metric induced from the usual metric on $\mathbb{R}$. Define the map $f : (a, b) \to (c, d)$ by the formula

$$\frac{f(x) - c}{x - a} = \frac{d - c}{b - a}.$$

We can represent this function as

$$f(x) = c + \frac{d - c}{b - a}(x - a).$$

By Proposition 3.1.25, f is continuous. Observe that f is bijective and the inverse map $f^{-1} : (c, d) \to (a, b)$ is given by

$$f^{-1}(y) = a + \frac{b - a}{d - c}(y - c).$$

By Proposition 3.1.25, f^{-1} is continuous. This shows that (a, b) is homeomorphic to (c, d).

Example 3.2.15. Consider the map f defined in Example 3.2.14. Note that $f(a) = c$ and $f(b) = d$. This shows that the closed intervals $[a, b]$ and $[c, d]$ are homeomorphic.

Example 3.2.16. Define the map $f : \mathbb{R} \to (-1, 1)$ by

$$f(x) = \frac{x}{1 + |x|}.$$

Consider the map $g : (-1, 1) \to \mathbb{R}$ given by

$$g(y) = \frac{y}{1 - |y|}.$$

Note that $g \circ f = I_{\mathbb{R}}$ and $f \circ g = I_{(-1,1)}$. The denominator of the value of f is the composition of two continuous maps $x \mapsto |x|$ and $x \mapsto 1 - x$. By Proposition 3.1.25(iii), f is continuous. Similarly, g is continuous. This shows that $(-1, 1)$ is homeomorphic to the real line $\mathbb{R}$.

Example 3.2.17. The map $f : (0, 1) \to (1, \infty)$ defined by $f(x) = \frac{1}{x}$ is a homeomorphism.

Example 3.2.18. Let a be a positive real number. Then the map $f : (1, \infty) \to (a, \infty)$ defined by $f(x) = ax$ is a homeomorphism.

Example 3.2.19. Let $a \in \mathbb{R}$. Then the map $f : (a, \infty) \to (-\infty, -a)$ defined by $f(x) = -x$ is a homeomorphism.

Example 3.2.20. Consider the usual norm on $\mathbb{R}^n$. Let r be a positive real number. Define the map $f : B(0, r) \to \mathbb{R}^n$ by

$$f(x) = \frac{x}{r - \|x\|}.$$

Note that f is continuous and the inverse map $f^{-1} : \mathbb{R}^n \to B(0, r)$ given by

$$f^{-1}(y) = \frac{ry}{1 + \|y\|}$$

is also continuous. This shows that the open ball $B(0, r)$ is homeomorphic to $\mathbb{R}^n$.

Example 3.2.21. Consider $A = \{(x, y) \in \mathbb{R}^2 \mid \max\{|x|, |y|\} = 1\}$ and $\mathbb{S}^1 = \{(x, y) \in \mathbb{R}^2 \mid x^2 + y^2 = 1\}$ as subspaces of $\mathbb{R}^2$. Define the map $f : A \to \mathbb{S}^1$ by

$$f(x, y) = \frac{(x, y)}{\sqrt{x^2 + y^2}}.$$

Note that the map $g : \mathbb{S}^1 \to A$ defined by

$$g(x, y) = \frac{(x, y)}{\max\{|x|, |y|\}}$$

is the inverse of f and f and g are continuous maps. Therefore A is homeomorphic to $\mathbb{S}^1$.

Theorem 3.2.22 (Tietze's extension theorem). *Let A be a closed set in a metric space, and let f : A → ℝ be a bounded continuous map. Then f can be extended to a bounded continuous map $\tilde{f} : X → ℝ$.*

Proof. If f is constant, then we can define $\tilde{f} : X → ℝ$ to be the same constant map. Suppose that f is not a constant map. Since f is bounded, the codomain of f is some closed interval $[u, v]$. Since closed intervals are homeomorphic to each other, we can compose a suitable homeomorphism to get the codomain $[1, 2]$. Therefore we can assume that f is a map from A to $[1, 2]$. Define the map $\tilde{f} : X → ℝ$ by

$$\tilde{f}(x) = \begin{cases} f(x) & \text{if } x \in A, \\ \frac{\inf\{f(y)d(x,y)|y \in A\}}{d(x,A)} & \text{if } x \in X \setminus A. \end{cases}$$

Note that $1 \le \tilde{f}(x) \le 2$ for all $x \in X$ and $\tilde{f}|_A = f$. We claim that $\tilde{f}$ is continuous on X. Let $\epsilon > 0$ and $a \in X$. Then either $a \in A$ or $a \in X \setminus A$. First, suppose that $a \in A$. Further, there may be two possibilities, either $a \in \text{Int} A$ or $a \in \text{Bd} A$. Suppose that $a \in \text{Int} A$. Then there exists $\delta_1 > 0$ such that $B(a, \delta_1) \subseteq \text{Int} A$. Note that $\tilde{f}(a) = f(a)$. Since f is continuous at a, there is an open ball $B(a, \delta_2)$ such that

$$f(B(a, \delta_2)) \subseteq (f(a) - \epsilon, f(a) + \epsilon).$$

Let $\delta = \min\{\delta_1, \delta_2\}$. Then

$$f(B(a, \delta)) \subseteq (f(a) - \epsilon, f(a) + \epsilon).$$

This implies that

$$\tilde{f}(B(a, \delta)) \subseteq (\tilde{f}(a) - \epsilon, \tilde{f}(a) + \epsilon).$$

Therefore $\tilde{f}$ is continuous at a.

Now suppose that $a \in \text{Bd} A \subseteq A$. Since f continuous at a, there is an open ball $B(a, \delta)$ in X such that

$$f(B(a, \delta) \cap A) \subseteq (f(a) - \epsilon, f(a) + \epsilon). \tag{3.4}$$

This implies that

$$\tilde{f}\left(B\left(a, \frac{\delta}{4}\right) \cap A\right) \subseteq (\tilde{f}(a) - \epsilon, \tilde{f}(a) + \epsilon). \tag{3.5}$$

Since a is a boundary point of A, $B(a, \frac{\delta}{4}) \cap (X \setminus A) \ne \emptyset$. Let $y \in B(a, \frac{\delta}{4}) \cap (X \setminus A)$. Now we claim that

$$\inf\{f(x)d(x,y) \mid x \in A\} = \inf\{f(x)d(x,y) \mid x \in B(a, \delta) \cap A\}. \tag{3.6}$$

If $x \notin B(a, \delta) \cap A$, then

$$d(x,y) \geq d(x,a) - d(a,y)$$
$$> \delta - \frac{\delta}{4}$$
$$= \frac{3\delta}{4}.$$

Since $f(x) \geq 1$ for all $x \in A$,

$$\inf\{f(x)d(x,y) \mid x \in B(a,\delta) \cap A\} \geq \frac{3\delta}{4}. \tag{3.7}$$

Also, note that $a \in B(a, \delta) \cap A$. Then

$$f(a)d(a,y) \leq 2d(a,y)$$
$$< 2\frac{\delta}{4}$$
$$< \frac{3\delta}{4}. \tag{3.8}$$

By equations (3.7) and (3.8), (3.6) holds. Similarly,

$$\inf\{d(x,y) \mid x \in B(x,\delta) \cap A\} = \inf\{d(x,y) \mid x \in A\}$$
$$= d(y,A). \tag{3.9}$$

By quation (3.4), for $x \in B(a, \delta) \cap A$, we have

$$f(a) - \epsilon < f(x) < f(a) + \epsilon.$$

This implies that

$$(f(a) - \epsilon)d(y,A) \leq \inf\{f(x)d(x,y) \mid x \in B(x,\delta) \cap A\}$$
$$\leq (f(a) + \epsilon)d(y,A).$$

By equation (3.6) we have

$$(f(a) - \epsilon)d(y,A) \leq \inf\{f(x)d(x,y) \mid x \in A\}$$
$$\leq (f(a) + \epsilon)d(y,A).$$

This implies that

$$(f(a) - \epsilon) \leq \frac{\inf\{f(x)d(x,y) \mid x \in A\}}{d(y,A)} \leq (f(a) + \epsilon).$$

Since $\tilde{f}(a) = f(a)$,

$$\tilde{f}(B(a,\delta) \cap (X \setminus A)) \subseteq (\tilde{f}(a) - \epsilon, \tilde{f}(a) + \epsilon). \tag{3.10}$$

By equations (3.5) and (3.10) we observe that $\tilde{f}$ is continuous at $a \in \operatorname{Bd} A$.

Finally, suppose that $a \in X \setminus A$. Define the map $g : X \setminus A \to \mathbb{R}$ by

$$g(y) = \inf\{f(x)d(x,y) \mid x \in A\}.$$

We claim that g is continuous at a. Take $y \in X \setminus A$ such that $d(y,a) < \frac{\epsilon}{4}$. Then for $x \in A$, we have

$$d(a,x) \leq d(a,y) + d(y,x) < \frac{\epsilon}{4} + d(y,x).$$

Then

$$f(x)d(a,x) < f(x)\frac{\epsilon}{4} + f(x)d(y,x) \leq \frac{\epsilon}{2} + f(x)d(y,x).$$

This implies that

$$\inf\{f(x)d(a,x) \mid x \in A\} \leq \frac{\epsilon}{2} + \inf\{f(x)d(x,y) \mid x \in A\}.$$

In other words,

$$g(a) \leq \frac{\epsilon}{2} + g(y).$$

Similarly,

$$g(y) \leq \frac{\epsilon}{2} + g(a).$$

This implies that

$$|g(y) - g(a)| \leq \frac{\epsilon}{2} < \epsilon.$$

Therefore g is continuous at a. Note that $\tilde{f}(x) = \frac{g(x)}{d(x,A)}$ on $X \setminus A$. Since $x \to d(x,A)$ is continuous and $d(x,A) > 0$, $\tilde{f}$ is continuous at $a \in X \setminus A$. $\square$

Remark 3.2.23. The above proof is due to Diedonne [2].

We can relax the condition of boundedness of the map in Tietze's extension theorem.

Proposition 3.2.24. *Let A be a closed set in a metric space, and let $f : A \to \mathbb{R}$ be a continuous map. Then f can be extended to a continuous map $\tilde{f} : X \to \mathbb{R}$.*

Proof. Let $a, b \in \mathbb{R}$, and let $\phi : \mathbb{R} \to (a,b)$ be a homeomorphism. Let $\psi = \phi \circ f$. Then $\psi : A \to (a,b) \subseteq [a,b]$ is a continuous map. By Tietze's extension theorem we have a continuous extension $\tilde{\psi} : X \to [a,b]$ of ψ. Let $B = \tilde{\psi}^{-1}\{a,b\}$. Then B is closed, and $B \cap A = \emptyset$.

By Urysohn's lemma there is a continuous map $g : X \to [0,1]$ such that $g(A) = 1$ and $g(B) = 0$. Let $\mu(x) = \tilde{\psi}(x)g(x)$. We can check that $x \to \mu(x)$ defines a continuous map from X to (a, b). Define the map $\tilde{f} : X \to \mathbb{R}$ by $\tilde{f}(x) = \phi^{-1}(\mu(x))$. Then $\tilde{f}$ is continuous, and for $x \in A$, we have

$$\begin{aligned}
\tilde{f}(x) &= \phi^{-1}(\mu(x)) \\
&= \phi^{-1}(\tilde{\psi}(x)g(x)) \\
&= \phi^{-1}(\tilde{\psi}(x)) \\
&= \phi^{-1}(\psi(x)) \\
&= f(x).
\end{aligned}$$

$\square$

Definition 3.2.25. A metric space X is said be homeomorphically embedded in a metric space Y if there is a map $f : X \to Y$ such that X is homeomorphic to the image $f(X)$ as a subspace of Y. Such a map f is called a homeomorphic embedding or simply embedding of X in Y.

Example 3.2.26. Let $\phi : (0,1) \to \mathbb{R}$ be a homeomorphism. Then the map $f : (0,1) \to \mathbb{R}^2$ defined by

$$f(x) = (\phi(x), 0)$$

is a homeomorphic embedding.

Definition 3.2.27. A point a in a metric space X is called an isolated point if there is a positive real number r such that

$$B(a, r) = \{r\}.$$

Example 3.2.28. In the discrete metric space, each point is an isolated point.

Example 3.2.29. Consider $A = (0,1) \cup \{2\}$ as a metric space of the real line. Then $\{2\}$ is the only isolated point of A.

The real line is a metric space without isolated points. Also, the sets of rational and irrational numbers are dense sets in the real line.

Proposition 3.2.30 (Sung Soo Kim). *Let X be a metric space without isolated points. Then X has a dense set whose complement is also dense.*

Proof. For given $\epsilon > 0$, consider a set S_ϵ in X satisfying the following two properties:
(A) For any two distinct points x and y of S_ϵ, $d(x, y) \geq \epsilon$,
(B) S_ϵ is maximal with respect to (A).

Since X has no isolated points, S_ϵ is nonempty. By Zorn's lemma we can observe that such a set exists for each $\epsilon > 0$.

We will inductively construct the disjoint sets $S_{\frac{1}{n}}$ for each $n \in \mathbb{N}$ satisfying (A) and (B). Consider a set S_1 with the above properties. Suppose we have found the disjoint sets

$$S_1, S_{\frac{1}{2}}, \dots, S_{\frac{1}{n}}$$

satisfying (A) and (B). Note that

$$S_1 \cup S_{\frac{1}{2}} \cup \dots \cup S_{\frac{1}{n}}$$

is nonempty and has no isolated point. Then there is a set $S_{\frac{1}{n+1}}$ satisfying the above properties that is disjoint from

$$S_1 \cup S_{\frac{1}{2}} \cup \dots \cup S_{\frac{1}{n}}.$$

Let $A = \bigcup_{n=1}^{\infty} S_{\frac{1}{2n}}$ and $B = \bigcup_{n=1}^{\infty} S_{\frac{1}{2n-1}}$. Observe that A and B are dense in X, $A \cap B = \emptyset$, and $A \cup B = X$. $\square$

The set $\mathbb{Q}$ of rational numbers as a subspace of the real line is a countable metric space without isolated points. We will show that such a metric space is unique up to homeomorphism. The following theorem was originally proved by Sierpinski, but we presentg the proof of Sierpinski's theorem given by F. K. Dashiell Jr. [1]. We recommend the reader to draw the figure while reading the proof of the following theorem. Let us first define the set W_G of all the words consisting of the symbols B and G such that the first letter is G. In other words,

$$W_G = \{G, GB, GG, GBB, GGB, \dots\}.$$

For $t \in W_G$, the length $|t|$ of the word t is defined as the number of letters appeared in t. For example, $|GBBGB| = 5$. Note that there are 2^{n-1} words of length n. We say that a word $t \in W_G$ is blue (respectively, green) if the last letter of t is B (respectively, G). For example, $t = GBBBGB$ is blue. By tG (respectively, tB) we mean that the word t is extended by adding the letter G (respectively, B) in the last. For example, if $t = GBGBBG$, then $tB = GBGBBGB$.

Theorem 3.2.31 (Sierpinski's theorem). *Let (X, d_X) and (Y, d_Y) be countable metric spaces without isolated points. Then X and Y are homeomorphic.*

Proof. Let

$$X = \{x_1, x_2, \dots\}$$

and

$$Y = \{y_1, y_2, \dots\}.$$

Let

$$D = \{d_X(x_n, x_m) \mid n \neq m\} \cup \{d_Y(y_n, y_m) \mid n \neq m\}.$$

For each nonempty set U of X (respectively, Y), define

$$\mu(U) = x_k \quad (\text{respectively}, \mu(U) = y_k),$$

where $k = \min\{i \mid x_i \in U\}$ (respectively, $k = \min\{i \mid y_i \in U\}$).

For each nonempty set U of X or Y and for $r > 0$, define the sets

$$B(U, r) = B(\mu(U), r) \quad \text{and} \quad G(U, r) = U \setminus B(U, r).$$

Let U be a nonempty open set of X. Let $x = \mu(U)$. Since x is not an isolated point, there is a point $y \in U$ such that $y \neq x$. Choose $0 < r < d_X(x, y)$ with $r \notin D$ such that

$$B(U, r) = B(x, r) \subseteq U.$$

Therefore $y \in U \setminus B(U, r) = G(U, r)$. This implies that $G(U, r)$ is nonempty. Since $r \notin D$,

$$G(U, r) = \{z \in U \mid d_X(x, z) > r\}.$$

Therefore $G(U, r)$ is open. Thus, for sufficiently small $r > 0$, $B(U, r)$ and $G(U, r)$ partition U into two nonempty open sets whenever U is open in X. A similar argument holds for an open set in Y.

Now we will describe a sequence of partitions of X and Y into nonempty open sets inductively on the length of words $t \in W_G$.

Let $U_G = X$ and $V_G = Y$. Using the mathematical induction on the length of $t \in W_G$, suppose that we have described the open sets U_t and V_t of X and Y, respectively, for all $t \in W_G$ with $|t| = n \geq 1$. Define

$$U_{tB} = B(U_t, r_{n+1}) \quad \text{and} \quad U_{tG} = G(U_t, r_{n+1}),$$

where $r_{n+1} < \frac{1}{n+1}$. Also, define

$$V_{tB} = B(V_t, r_{n+1}) \quad \text{and} \quad V_{tG} = G(V_t, r_{n+1}),$$

where $r_{n+1} < \frac{1}{n+1}$.

Suppose that $x \in X$ is such that $x = \mu(U_t)$ for some green $t \in W_G$. Then x becomes the center of the open balls U_s for all $s \in \{tB, tBB, \dots\}$. Therefore we have a decreasing sequence of open balls centered at x with the radii tending to zero. This shows that each $x \in X$ is associated with a unique green $t \in W_G$. We can similarly argue for each $y \in Y$. Therefore this defines the bijective map $f : X \to Y$ by

$$f(\mu(U_t)) = \mu(V_t) \quad \text{for green } t \in W_G.$$

Note that $f(U_t) = V_t$ for all $t \in W_G$. Let $f(x) = y$. Let $x \in U_t$ and $y \in V_t$, where $t \in W_G$ is blue. Then for $\epsilon > 0$, we can find $\delta > 0$ such that

$$f(B(x, \delta)) \subseteq B(y, \epsilon).$$

This shows that f is continuous. We can similarly show that $f^{-1} : Y \to X$ is also continuous. Hence f is a homeomorphism. □

Corollary 3.2.32. *Let (X, d_X) be a countable metric space. Then X is homeomorphically embedded in the subspace $\mathbb{Q}$ of the real line.*

Proof. Consider $X \times \mathbb{Q}$ with the metric

$$d((x, r), (y, s)) = d_X(x, y) + |r - s|.$$

Since $X \times \mathbb{Q}$ has no isolated points, by Theorem 3.2.31, $X \times \mathbb{Q}$ is homeomorphic to $\mathbb{Q}$. Note that X is homeomorphically embedded in $X \times \mathbb{Q}$. Therefore X is homeomorphically embedded in $\mathbb{Q}$. □

For a metric space X, let $\mathrm{Homeo}(X)$ denote the set of all homeomorphisms from X to X. We can check that $\mathrm{Homeo}(X)$ is a group with respect to the composition of maps.

Proposition 3.2.33. *Let (a, b) and $[a, b]$ be the metric spaces considered as the subspaces of the real line. Then the group $\mathrm{Homeo}([a, b])$ is isomorphic to the group $\mathrm{Homeo}((a, b))$.*

Proof. Let $f \in \mathrm{Homeo}([a, b])$. Then we can check that the restriction of f on (a, b) is a member of $\mathrm{Homeo}((a, b))$. Conversely, every member $f \in \mathrm{Homeo}((a, b))$ can be uniquely extended to a homeomorphism $\bar{f} : [a, b] \to [a, b]$ defined by

$$\bar{f}(x) = \begin{cases} f(x) & \text{if } x \in (a, b), \\ \lim_{x \to a} f(x) & \text{if } x = a, \\ \lim_{x \to b} f(x) & \text{if } x = b. \end{cases}$$

Now we can easily check that the group $\mathrm{Homeo}([a, b])$ is isomorphic to the group $\mathrm{Homeo}((a, b))$. □

Define $\tilde{d} : \mathrm{Homeo}([a, b]) \times \mathrm{Homeo}([a, b]) \to \mathbb{R}$ by

$$\tilde{d}(f, g) = \sup\{|f(x) - g(x)|, |f^{-1}(x) - g^{-1}(x)| \mid x \in [a, b]\}.$$

We claim that $\tilde{d}$ is a metric on $\mathrm{Homeo}([a, b])$. We will only prove the triangle inequality, and the other conditions are left as an exercise. Let $f, g, h \in \mathrm{Homeo}([a, b])$. Choose $\epsilon > 0$. Then there is a point $x \in [a, b]$ such that

$$\tilde{d}(f,g) \le |f(x) - g(x)| + \epsilon$$

or

$$\tilde{d}(f,g) \le |f^{-1}(x) - g^{-1}(x)| + \epsilon.$$

Then

$$
\begin{aligned}
\tilde{d}(f,g) &\le |f(x) - g(x)| + \epsilon \\
&\le |f(x) - h(x)| + |h(x) - g(x)| + \epsilon \\
&\le \tilde{d}(f,h) + \tilde{d}(h,g) + \epsilon,
\end{aligned}
\tag{3.11}
$$

or

$$
\begin{aligned}
\tilde{d}(f,g) &\le |f^{-1}(x) - g^{-1}(x)| + \epsilon \\
&\le |f^{-1}(x) - h^{-1}(x)| + |h^{-1}(x) - g^{-1}(x)| + \epsilon \\
&\le \tilde{d}(f,h) + \tilde{d}(h,g) + \epsilon.
\end{aligned}
\tag{3.12}
$$

Since $\epsilon > 0$ is arbitrary, by (3.11) and (3.12) we have

$$\tilde{d}(f,g) \le \tilde{d}(f,h) + \tilde{d}(h,g).$$

The mathematicians have tried to understand the structure of the group Homeo($[a,b]$). As an example, we give the following result by Fine and Schweigert [3]. We will not give its proof. We recommend the author to go through [3] for details.

Theorem 3.2.34 (Fine and Schweigert). *Every element of* Homeo($[a,b]$) *may be written as a product of at most four involutions, and the number four is best possible.*

Note that the map on Homeo($[a,b]$) sending an element to its inverse is continuous. After we define the product metric, we can prove that the binary operation on Homeo($[a,b]$) is also continuous.

3.3 Equivalent metrics

Definition 3.3.1. The metrics d_1 and d_2 on a set X are called Lipschitz equivalent if there are positive real numbers m and M such that

$$m\, d_1(x,y) \le d_2(x,y) \le M\, d_1(x,y)$$

for all $x,y \in X$.

Example 3.3.2. Consider the metrics d_p and d_∞ on $\mathbb{R}^n$ as defined in Examples 2.1.3 and 2.1.4. Let

$$x = (x_1, \ldots, x_n) \quad \text{and} \quad y = (y_1, \ldots, y_n) \quad \text{in } \mathbb{R}^n.$$

Since

$$d_p(x,y) \geq |x_i - y_i| \quad \text{for all } i = 1, \ldots, n,$$
$$d_p(x,y) \geq \max\{|x_i - y_i| \mid 1 \leq i \leq n\} = d_\infty(x,y).$$

Also,

$$d_p(x,y) = \left(\sum_{i=1}^{n} |x_i - y_i|^p \right)^{\frac{1}{p}}$$

$$\leq \left(\sum_{i=1}^{n} d_\infty(x,y)^p \right)^{\frac{1}{p}}$$

$$= n^{\frac{1}{p}} d_\infty(x,y).$$

Therefore

$$d_\infty(x,y) \leq d_p(x,y) \leq n^{\frac{1}{p}} d_\infty(x,y).$$

Hence, for each $p \geq 1$, d_p and d_∞ are Lipschitz equivalent.

Example 3.3.3. Let $\mathbb{R}$ be the real line. Let $\bar{d}$ be the metric on $\mathbb{R}$ as defined in Example 2.1.9. Then the usual metric on $\mathbb{R}$ is not Lipschitz equivalent to $\bar{d}$.

Definition 3.3.4. The metrics d_1 and d_2 on a set X are called topologically equivalent if U is open in (X, d_1) if and only if U is open in (X, d_2).

Let τ_1 and τ_2 be the collection of all open sets in (X, d_1) and (X, d_2), respectively. Then metrics d_1 and d_2 are topologically equivalent if $\tau_1 = \tau_2$. We say that two metric spaces (X, d_1) and (X, d_2) are topologically equivalent if d_1 and d_2 are topologically equivalent. Observe that the relations of being Lipschitz equivalent and topologically equivalent are equivalence relations on a set of metrics on a set X. Also observe that metric spaces (X, d_1) and (X, d_2) are topologically equivalent if and only if the identity map $I : (X, d_1) \rightarrow (X, d_2)$ is a homeomorphism. The following is an easy observation. Hence its proof is left as an exercise.

Proposition 3.3.5. *A metric space (X, d_1) is topologically equivalent to a metric space (X, d_2) if and only if for each open ball $B_{d_1}(x, \epsilon)$ in (X, d_1), there is an open ball $B_{d_2}(x, \delta)$ in (X, d_2) such that*

$$B_{d_2}(x, \delta) \subseteq B_{d_1}(x, \epsilon),$$

and for each open ball $B_{d_2}(x, \epsilon')$ in (X, d_2), there is an open ball $B_{d_1}(x, \delta')$ in (X, d_1) such that

$$B_{d_1}(x, \delta') \subseteq B_{d_2}(x, \epsilon').$$

Proposition 3.3.6. *If two metrics are Lipschitz equivalent, then they are topologically equivalent.*

Proof. Let d_1 and d_2 be two Lipschitz equivalent metrics on a set X. Then there are positive real numbers m and M such that

$$m \, d_1(x, y) \le d_2(x, y) \le M \, d_1(x, y)$$

for all $x, y \in X$. Therefore for all $x \in X$ and $\epsilon > 0$, we have

$$B_{d_1}(x, \epsilon) \subseteq B_{d_2}(x, M\epsilon) \quad \text{and} \quad B_{d_2}(x, m\epsilon) \subseteq B_{d_1}(x, \epsilon).$$

Hence, by Proposition 3.3.5, d_1 is topologically equivalent to d_2. $\qquad\square$

Example 3.3.7. Let d_1 be the discrete metric on $\mathbb{N}$, and let d_2 be the metric on $\mathbb{N}$ induced by the usual metric on $\mathbb{R}$. Then d_1 is topologically equivalent to d_2 on $\mathbb{N}$, but they are not Lipschitz equivalent.

Proposition 3.3.8. *Let (X, d) be a metric space, and let*

$$\overline{d}(x, y) = \min\{1, d(x, y)\}$$

for $x, y \in X$. Then d and $\overline{d}$ are topologically equivalent.

Proof. Since $\overline{d}(x, y) \le d(x, y)$ for all $x, y \in X$, for each $\epsilon > 0$ and $x \in X$, we have

$$B_d(x, \epsilon) \subseteq B_{\overline{d}}(x, \epsilon).$$

Now, consider an open ball $B_d(x, \epsilon)$ in the metric space (X, d). Let $\delta = \min\{1, \epsilon\}$. Note that if $\overline{d}(x, y) < \delta$, then $\overline{d}(x, y) < 1$. Therefore $d(x, y) = \overline{d}(x, y) < \delta \le \epsilon$. This shows that if $y \in B_{\overline{d}}(x, \delta)$, then $y \in B_d(x, \epsilon)$. Hence

$$B_{\overline{d}}(x, \delta) \subseteq B_d(x, \epsilon).$$

Thus, by Proposition 3.3.5, d and $\overline{d}$ are topologically equivalent. $\qquad\square$

Corollary 3.3.9. *Each metric on a set X is topologically equivalent to a bounded metric on X.*

Let $\| \cdot \|_1$ and $\| \cdot \|_2$ be two norms on a vector space V. Let d_1 and d_2 be metrics on V induced by $\| \cdot \|_1$ and $\| \cdot \|_2$, respectively. We say that the norms $\| \cdot \|_1$ and $\| \cdot \|_2$ on V are Lipschitz equivalent (respectively, topologically equivalent) if d_1 and d_2 are Lipschitz equivalent (respectively, topologically equivalent). Clearly, $\| \cdot \|_1$ and $\| \cdot \|_2$ are Lipschitz equivalent (respectively, topologically equivalent) if there are positive real numbers m and M such that

$$m \|x\|_1 \leq \|x\|_2 \leq M \|x\|_1$$

for all $x \in V$.

Proposition 3.3.10. *Any two norms $\| \cdot \|_1$ and $\| \cdot \|_2$ on a vector space V are Lipschitz equivalent if they are topologically equivalent.*

Proof. By Proposition 3.3.6 Lipschitz equivalent norms are topologically equivalent. Conversely, suppose that $\| \cdot \|_1$ and $\| \cdot \|_2$ are topologically equivalent on a vector space V. Then for each open ball $B_{d_2}(x, \epsilon)$, there is an open ball $B_{d_1}(x, \delta)$ such that

$$B_{d_1}(x, \delta) \subseteq B_{d_2}(x, \epsilon).$$

This implies that

$$x + B_{d_1}(0, \delta) \subseteq x + B_{d_2}(0, \epsilon).$$

Equivalently, we have

$$B_{d_1}(0, 1) \subseteq B_{d_2}\left(0, \frac{\epsilon}{\delta}\right).$$

We claim that for all $x \in X$,

$$\|x\|_2 \leq \frac{\epsilon}{\delta} \|x\|_1.$$

On the contrary, suppose that there is $a \in X$ such that

$$\|a\|_2 > \frac{\epsilon}{\delta} \|a\|_1.$$

Then there is a real number $k > 1$ such that

$$\|a\|_2 = k \frac{\epsilon}{\delta} \|a\|_1.$$

Note that $a \neq 0$; otherwise, $0 > 0$, which is a contradiction. Let $y = \frac{a}{\|a\|_1}$. Then

$$\left\| \frac{y}{k} \right\|_2 = \frac{\epsilon}{\delta}.$$

Also, note that

$$\left\|\frac{y}{k}\right\|_1 = \frac{1}{k} < 1.$$

This shows that $\frac{y}{k} \in B_{d_1}(0,1)$ but $\frac{y}{k} \notin B_{d_2}(0,\frac{\epsilon}{\delta})$. This is a contradiction.

Again by the topological equivalence of the norms $\|\cdot\|_1$ and $\|\cdot\|_2$, for each open ball $B_{d_1}(x,\epsilon')$, there is an open ball $B_{d_2}(x,\delta')$ such that

$$B_{d_2}(x,\delta') \subseteq B_{d_1}(x,\epsilon').$$

By a similar argument, for all $x \in X$, we have

$$\|x\|_1 \le \frac{\epsilon'}{\delta'}\|x\|_2.$$

This implies that

$$\frac{\delta'}{\epsilon'}\|x\|_1 \le \|x\|_2 \le \frac{\epsilon}{\delta}\|x\|_1.$$

Hence $\|\cdot\|_1$ and $\|\cdot\|_2$ are Lipschitz equivalent. $\qquad\qquad\square$

Example 3.3.11. Consider the norms $\|\cdot\|$ and $\|\cdot\|_\infty$ on $C[0,1]$ as defined in Examples 2.2.8 and 2.2.9. Then $\|\cdot\|$ and $\|\cdot\|_\infty$ are not Lipschitz equivalent on $C[0,1]$. On the contrary, suppose that $\|\cdot\|$ and $\|\cdot\|_\infty$ are Lipschitz equivalent on $C[0,1]$. Then for all $f \in C[0,1]$, there are positive real numbers m and M such that

$$m\|f\| \le \|f\|_\infty \le M\|f\|.$$

In particular, take $f_n(x) = x^n$, where $n \in \mathbb{N}$. Then we have

$$\frac{m}{n+1} \le 1 \le \frac{M}{n+1}.$$

This is a contradiction.

Example 3.3.12. Consider a norm $\|\cdot\|$ on an infinite-dimensional vector space V. Let $f : V \to \mathbb{F}$ be a discontinuous linear functional, where $\mathbb{F} = \mathbb{R}$ or $\mathbb{F} = \mathbb{C}$. Define a new norm $\|\cdot\|_f$ on V by

$$\|x\|_f = \|x\| + |f(x)|.$$

Then $\|\cdot\|$ and $\|\cdot\|_f$ are not Lipschitz equivalent on V. Indeed, suppose, on the contrary, that $\|\cdot\|$ and $\|\cdot\|_f$ are Lipschitz equivalent on V. Then for all $x \in V$, there are positive real numbers m and M such that

$$m\|x\| \le \|x\|_f \le M\|x\|.$$

This implies that for all $x \in V$, we have

$$(m-1)\|x\| \le |f(x)| \le (M-1)\|x\|.$$

By Proposition 3.1.32, f is continuous. This is a contradiction.

Proposition 3.3.13. *Let (X,d) be a metric space, and let $f : [0,\infty) \to [0,\infty)$ be continuous, subadditive, amenable, and increasing map. Then d and $f \circ d$ are topologically equivalent.*

Proof. Since f is continuous at 0, for each $\epsilon > 0$, there is $\delta > 0$ such that

$$0 \le a\delta \Rightarrow f(a) < \epsilon.$$

Taking $a = d(x,y)$, we get

$$d(x,y) < \delta \Rightarrow f \circ d(x,y) < \epsilon.$$

This shows that

$$B_d(x,\delta) \subseteq B_{f \circ d}(x,\epsilon).$$

Now let $a, b \in [0,\infty)$ be such that $a \le 2b$. Then

$$f(a) \le f(2b) \le 2f(b).$$

This implies that if

$$f(a) > 2f(b),$$

then

$$a > 2b.$$

Let $\epsilon' > 0$. Then taking $a = \epsilon'$ and $b = d(x,y)$, we get

$$f(d(x,y)) < \frac{f(\epsilon')}{2} \Rightarrow d(x,y) < \frac{\epsilon'}{2} < \epsilon'.$$

This shows that

$$B_{f \circ d}\left(x, \frac{f(\epsilon')}{2}\right) \subseteq B_d(x, \epsilon').$$

Thus d and $f \circ d$ are topologically equivalent. ☐

Example 3.3.14. Let $f : [0, \infty) \to [0, \infty)$ be the map defined by

$$f(x) = \frac{x}{1 + x}.$$

Note that f satisfies the conditions of Proposition 3.3.13. Therefore the metrics d and $\frac{d}{1+d}$ are topologically equivalent.

Proposition 3.3.15. *Let $f : [0, \infty) \to [0, \infty)$ be a metric-preserving map. Then $f \circ d$ is topologically equivalent to the discrete metric for every metric d if and only if f is discontinuous at 0.*

Proof. Suppose that f is discontinuous at 0. Let (X, d) be a metric space, and let $x \in X$. By Proposition 2.3.13 there is $\epsilon > 0$ such that

$$f(y) \geq \epsilon \quad \text{for all } y > 0.$$

This implies that

$$B_{f \circ d}(x, \epsilon) = \{x\}.$$

Hence $f \circ d$ is topologically equivalent to the discrete metric.

Conversely, suppose that $f \circ d$ is topologically equivalent to the discrete metric for every metric d. Let d be the usual metric on $\mathbb{R}$. Let $\epsilon > 0$ be such that

$$B_{f \circ d}(0, \epsilon) = \{0\}.$$

Let δ be an arbitrary positive real number. Choose $0 \leq x_1, x_2 < \delta$ such that $f(x_1) \neq f(x_2)$. Then one of $f(x_1)$ or $f(x_2)$ will not satisfy the condition $f(x) < \epsilon$, since otherwise

$$f(x_1), f(x_2) \in B_{f \circ d}(f(0), \epsilon) = B_{f \circ d}(0, \epsilon) = \{0\},$$

a contradiction. Hence f is discontinuous at 0. $\square$

Proposition 3.3.16 (Borsik and Dobos). *Let $f : [0, \infty) \to [0, \infty)$ be a metric-preserving map. Then d and $f \circ d$ are topologically equivalent for every metric d if and only if f is continuous at 0.*

Proof. Suppose that d and $f \circ d$ are topologically equivalent for every metric d. Let d_u be the usual metric on $\mathbb{R}$. On the contrary, suppose that f is discontinuous at 0. By Proposition 3.3.15, $f \circ d_u$ is topologically equivalent to the discrete metric. This shows that d_u is topologically equivalent to the discrete metric. This is a contradiction.

Conversely, suppose that f is continuous at 0. Let (X, d) be a metric space, and let $x \in X$. Let $\epsilon > 0$. Since f is continuous at 0, we can choose $\delta \leq \epsilon$ such that

$$0 \leq y < \delta \Rightarrow f(y) < \epsilon.$$

This implies that

$$B_d(x, \delta) \subseteq B_{f \circ d}(x, \epsilon).$$

Now let $r > 0$. By Proposition 2.3.14 there is $\epsilon > 0$ such that

$$y \geq r \Rightarrow f(y) \geq \epsilon.$$

This implies that

$$B_{f \circ d}(x, \epsilon) \subseteq B_d(x, r).$$

Therefore d and $f \circ d$ are topologically equivalent. □

Let x and y be nonzero points in $\mathbb{R}^n$. Let $\langle x, y \rangle$ be the usual inner product on $\mathbb{R}^n$. By the Cauchy–Schwarz inequality we have

$$-1 \leq \frac{\langle x, y \rangle}{\|x\|\|y\|} \leq 1.$$

Since the cosine is continuous and strictly decreasing on $[0, \pi]$, there is a unique angle $\theta(x, y)$ with $0 \leq \theta(x, y) \leq \pi$ such that

$$\cos \theta(x, y) = \frac{\langle x, y \rangle}{\|x\|\|y\|}.$$

Consider the set

$$\mathbb{S}^n = \{x \in \mathbb{R}^{n+1} \mid \|x\| = 1\}.$$

This set is called the *n-sphere* or the *unit sphere of dimension n*.

Proposition 3.3.17. *The map* $d_{\mathbb{S}^n} : \mathbb{S}^n \times \mathbb{S}^n \to [0, \pi] \subseteq \mathbb{R}$ *defined by*

$$d_{\mathbb{S}^n}(x, y) = \theta(x, y),$$

where $\cos \theta(x, y) = \langle x, y \rangle$, *is a metric on* $\mathbb{S}^n$.

Proof. By the definition, $d_{\mathbb{S}^n}(x, y) \geq 0$. Clearly,

$$d_{\mathbb{S}^n}(x, y) = 0 \iff \langle x, y \rangle = 1.$$

Hence the equality holds in the Cauchy–Schwarz inequality. This implies that there is $\alpha \in \mathbb{R}$ such that $x = \alpha y$. Since $\|x\| = \|y\| = 1$, $\alpha = \pm 1$. Also, since $\langle x, y \rangle = 1$, $\alpha = 1$. This shows that

$$d_{\mathbb{S}^n}(x, y) = 0 \iff x = y.$$

Now since $\langle x, y \rangle = \langle y, x \rangle$ and the cosine is injective on $[0, \pi]$,

$$d_{\mathbb{S}^n}(x, y) = d_{\mathbb{S}^n}(y, x).$$

Now we will prove the triangle inequality. Let $x, y, z \in \mathbb{S}^n$. Then the dimension of the subspace generated by x, y, and z in $\mathbb{R}^{n+1}$ is not greater than 3. By the Gram–Schmidt process we get an orthogonal linear transformation U such that

$$U(x) = (1, 0, 0, \ldots, 0), \quad U(x) = (0, 1, 0, \ldots, 0), \quad \text{and} \quad U(z) = (0, 0, 1, 0, \ldots, 0).$$

Therefore, without loss of generality, we may assume that $n = 2$, that is, we can perform all the calculations in $\mathbb{R}^3$. Suppose that for $x, y \in \mathbb{R}^3$, $x \times y$ denotes the usual cross product. Then

$$
\begin{aligned}
\cos(\theta(x, y) + \theta(y, z)) &= \cos \theta(x, y) \cos \theta(y, z) - \sin \theta(x, y) \sin \theta(y, z) \\
&= \langle x, y \rangle \langle y, z \rangle - \|x \times y\| \|y \times z\| \\
&\leq \langle x, y \rangle \langle y, z \rangle - \langle x \times y, y \times z \rangle \\
&= \langle x, y \rangle \langle y, z \rangle - \langle x, y \rangle \langle y, z \rangle + \langle x, z \rangle \langle y, y \rangle \\
&= \langle x, z \rangle = \cos \theta(x, z).
\end{aligned}
$$

Since the cosine is decreasing on $[0, \pi]$,

$$\theta(x, z) \leq \theta(x, y) + \theta(y, z).$$

In other words,

$$d_{\mathbb{S}^n}(x, z) \leq d_{\mathbb{S}^n}(x, y) + d_{\mathbb{S}^n}(y, z). \qquad \square$$

Observe that $d_{\mathbb{S}^n}(x, y)$ is the arc length from x to y on the sphere traveled on the greater circle (see Figure 3.1).

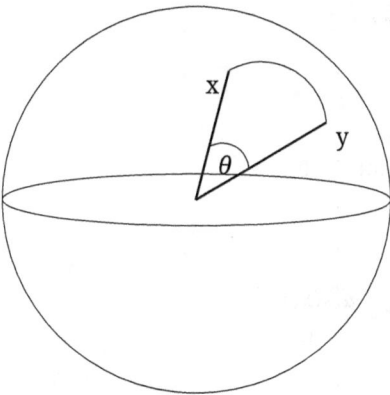

Figure 3.1: The distance $d_{\mathbb{S}^n}(x, y)$.

For $x, y \in \mathbb{S}^n \subseteq \mathbb{R}^{n+1}$, let $d_{\mathbb{R}^{n+1}}(x, y)$ denote the distance between x and y induced by the usual metric on $\mathbb{R}^{n+1}$.

Proposition 3.3.18. *The metric $d_{\mathbb{S}^n}$ and the induced Euclidean metric $d_{\mathbb{R}^{n+1}}$ on $\mathbb{S}^n$ are Lipschitz equivalent.*

Proof. Note that $\sin t \le t$ for $t \ge 0$. Then for $x, y \in Sp^n$, we have

$$
\begin{aligned}
d_{\mathbb{R}^{n+1}}(x, y)^2 &= \|x - y\|^2 \\
&= \|x\|^2 - 2\langle x, y\rangle + \|y\|^2 \\
&= 2 - 2\langle x, y\rangle \\
&= 2 - 2\cos d_{\mathbb{S}^2}(x, y) \\
&= 4\sin^2\left(\frac{d_{\mathbb{S}^n}(x, y)}{2}\right) \\
&\le 4\left(\frac{d_{\mathbb{S}^n}(x, y)}{2}\right)^2 \\
&= d_{\mathbb{S}^n}(x, y)^2.
\end{aligned}
$$

Therefore

$$
d_{\mathbb{R}^{n+1}}(x, y) \le d_{\mathbb{S}^n}(x, y).
$$

Again, note that the derivative of

$$
\phi(t) = \cos t - 1 + \frac{t^2}{2}
$$

is 0 at $t = 0$ and positive for $t > 0$. This shows that $\phi(t)$ is increasing for $t \ge 0$ and $\phi(t) \ge 0$ for $t \ge 0$. Using this fact and a similar argument, we get

$$
\sin t - t + \frac{t^3}{6} \ge 0 \quad \text{for } t \ge 0.
$$

This in turn implies that

$$
1 - \cos t - \frac{t^2}{2} + \frac{t^4}{24} \ge 0 \quad \text{for } t \ge 0.
$$

Putting $t = d_{\mathbb{S}^n}(x, y)$ above, we get

$$
2 - 2\cos d_{\mathbb{S}^n}(x, y) \ge d_{\mathbb{S}^n}(x, y)^2 - \frac{d_{\mathbb{S}^n}(x, y)^4}{12}.
$$

Hence

$$
\begin{aligned}
d_{\mathbb{R}^{n+1}}(x,y) &= \|x-y\|^2 \\
&= \|x\|^2 - 2\langle x,y\rangle + \|y\|^2 \\
&= 2 - 2\langle x,y\rangle \\
&= 2 - 2\cos d_{\mathbb{S}^2}(x,y) \\
&\geq d_{\mathbb{S}^n}(x,y)^2 - \frac{d_{\mathbb{S}^n}(x,y)^4}{12} \\
&\geq \left(1 - \frac{\pi^2}{12}\right) d_{\mathbb{S}^n}(x,y)^2 \quad (\text{as } d_{\mathbb{S}^n}(x,y) \leq \pi).
\end{aligned}
$$

This shows that $d_{\mathbb{S}^n}$ and $d_{\mathbb{R}^{n+1}}$ on $\mathbb{S}^n$ are Lipschitz equivalent. $\square$

Corollary 3.3.19. *The metric $d_{\mathbb{S}^n}$ and the induced Euclidean metric $d_{\mathbb{R}^{n+1}}$ on $\mathbb{S}^n$ are topologically equivalent.*

Remark 3.3.20. The geometries of these two metric spaces are different. This topic is out of the scope, so we will not discuss it here. As a general remark, we would like to mention that the sum of internal angles of a triangle in such geometry other than Euclidean geometry is not $180°$.

3.4 Isometry

Definition 3.4.1. The metric spaces (X, d_X) and (Y, d_Y) are called metrically equivalent or isometric if there exists a surjective map $f : X \to Y$ such that

$$
d_X(x,y) = d_Y\big(f(x), f(y)\big)
$$

for all $x, y \in X$. Such a map f is called an isometry between X and Y.

Note that an isometry is always injective and a surjective isometric embedding is an isometry. Also, note that the relation of being isometric is an equivalence relation on a set of metric spaces. By Example 2.1.15 we observe that if (X, d) is a metric space and Y is a set that is in the bijective correspondence with X, then we can define a metric on Y such that X is isometric to Y.

Example 3.4.2. Define the map $f : (0, \pi) \to \mathbb{S}^1$ by

$$
f(\theta) = (\cos\theta, \sin\theta).
$$

Clearly, f is injective. Let $\theta_1, \theta_2 \in (0,\pi)$ with $\theta_1 < \theta_2$. Let $x = (\cos\theta_1, \sin\theta_1)$ and $y = (\cos\theta_2, \sin\theta_2)$. Then

$$
\begin{aligned}
\cos d_{\mathbb{S}^1}(x,y) &= \langle x, y \rangle \\
&= \cos\theta_1 \cos\theta_2 + \sin\theta_1 \sin\theta_2 \\
&= \cos(\theta_2 - \theta_1).
\end{aligned}
$$

Since cosine is a bijection from $(0,\pi)$ to $(-1,1)$,

$$
d_{\mathbb{S}^1}(x,y) = \theta_2 - \theta_1 = |\theta_2 - \theta_1|.
$$

This shows that $(0,\pi)$ with the metric induced by the usual metric on $\mathbb{R}$ is isometrically embedded in $(\mathbb{S}^1, d_{\mathbb{S}^1})$.

Example 3.4.3. Define the map $d_l : (0,\infty) \times (0,\infty) \to \mathbb{R}$ by

$$
d_l(x,y) = |\log x - \log y|.
$$

We can check that d_l is a metric on $(0,\infty)$ and the map $f : (0,\infty) \to \mathbb{R}$ defined by $f(x) = \log x$ is an isometry.

Example 3.4.4. Let d_1 and d_2 be metrics on $\mathbb{N}$ as defined in Example 3.3.7. Then the identity map on $\mathbb{N}$ is not an isometry.

For a metric space (X, d), let $\mathrm{Iso}(X, d)$ denote the set of all isometries on X. When the metric d is given, we denote it by $\mathrm{Iso}(X)$. Then $\mathrm{Iso}(X)$ forms a group under the composition of maps. We now find the group $\mathrm{Iso}(\mathbb{R}^n)$ of the isometries of the Euclidean space $\mathbb{R}^n$.

Let A be an $n \times n$ real matrix, and let $b \in \mathbb{R}^n$. Viewing the element $x \in \mathbb{R}^n$ as a row matrix, we define the map $f_A^b : \mathbb{R}^n \to \mathbb{R}^n$ by $f_A^b(x) = Ax^t + b$, where x^t denotes the transpose of x. Note that the map f_A^b is uniquely determined by A and b. Indeed, let $f_{A_1}^{b_1} = f_{A_2}^{b_2}$. Then for all $x \in \mathbb{R}^n$, we have

$$
A_1 x^t + b_1 = A_2 x^t + b_2.
$$

This implies that for all $x \in \mathbb{R}^n$, we have

$$
(A_1 - A_2)x^t = b_2 - b_1.
$$

Hence $A_1 = A_2$ and $b_1 = b_2$. An $n \times n$ real matrix A is called an orthogonal matrix if $A^t A = I$, where A^t is the transpose of A.

Proposition 3.4.5. *Let A be an $n \times n$ real matrix, and let $b \in \mathbb{R}^n$. The map $f_A^b : \mathbb{R}^n \to \mathbb{R}^n$ by $f_A^b(x) = Ax^t + b$ is an isometry if and only if A is an orthogonal matrix.*

Proof. Note that f_A^b is an isometry if and only if the map $\phi : \mathbb{R}^n \to \mathbb{R}^n$ defined by $\phi(x) = Ax^t$ is an isometry. Also, ϕ is an isometry if and only if

$$\|\phi(x)\| = \|x\|$$

for all $x \in \mathbb{R}^n$. Now

$$
\begin{aligned}
\|\phi(x)\| = \|x\| &\iff \|\phi(x)\|^2 = \|x\|^2 \\
&\iff \langle \phi(x), \phi(x) \rangle = \langle x, x \rangle \\
&\iff xA^tAx^t = xx^t \\
&\iff x(A^tA - I)x^t = 0.
\end{aligned}
$$

Therefore f_A^b is an isometry if and only if A is an orthogonal matrix. $\square$

Proposition 3.4.6. *Let $f : \mathbb{R}^n \to \mathbb{R}^n$. Then the following statements are equivalent:*
(i) *f is an isometry such that $f(0) = 0$;*
(ii) *f preserves the inner product, that is, for all $x, y \in \mathbb{R}^n$,*

$$\langle f(x), f(y) \rangle = \langle x, y \rangle;$$

(iii) *f is linear transformation such that $f(x) = Ax^t$, where A is an orthogonal matrix.*

Proof. Suppose that (i) holds. Let $x, y \in \mathbb{R}^n$. Then

$$\|f(x) - f(y)\| = \|x - y\|.$$

Putting $y = 0$, for all $x \in \mathbb{R}^n$, we have

$$\|f(x)\| = \|x\|.$$

Now

$$
\begin{aligned}
\langle f(x) - f(y), f(x) - f(y) \rangle &= \|f(x) - f(y)\|^2 \\
&= \|x - y\|^2 \\
&= \langle x - y, x - y \rangle.
\end{aligned}
$$

This implies that

$$\langle f(x), f(y) \rangle = \langle x, y \rangle. \tag{3.13}$$

Now suppose that (ii) holds. Let $x, y, z \in \mathbb{R}^n$. Then

$$\langle f(x+y), f(z) \rangle = \langle x+y, z \rangle$$
$$= \langle x, z \rangle + \langle y, z \rangle$$
$$= \langle f(x), f(z) \rangle + \langle f(y), f(z) \rangle$$
$$= \langle f(x) + f(y), f(z) \rangle. \tag{3.14}$$

By equation (3.13) we can observe that

$$\{f(e_1), \ldots, f(e_n)\}$$

is an orthonormal basis of $\mathbb{R}^n$, where $\{e_1, \ldots, e_n\}$ is the standard basis of $\mathbb{R}^n$. Hence, taking $z = e_1, \ldots, e_n$ in equation (3.14), we get

$$f(x+y) = f(x) + f(y).$$

Let $a \in \mathbb{R}$. Then

$$\langle f(ax), f(y) \rangle = \langle ax, y \rangle$$
$$= a\langle x, y \rangle$$
$$= a\langle f(x), f(y) \rangle$$
$$= \langle af(x), f(y) \rangle. \tag{3.15}$$

By a similar argument we get

$$f(ax) = af(x).$$

Therefore f is a linear transformation. Hence there exists an $n \times n$ real matrix A such that $f(x) = Ax^t$. By the assumption, for all $x \in \mathbb{R}^n$,

$$\langle Ax^t, Ax^t \rangle = \langle x, x \rangle.$$

This implies that for all $x \in \mathbb{R}^n$, we have

$$x(A^t A - I)x^t = 0.$$

Therefore A is an orthogonal matrix.

Suppose that (iii) holds. Let $f(x) = Ax^t$, where A is an orthogonal matrix. Then, clearly, $f(0) = 0$. By a similar argument as in the proof of Proposition 3.4.5, f is an isometry. □

For $a \in \mathbb{R}^n$, let t_a denote a map on $\mathbb{R}^n$ defined by $t_a(x) = x + a$. Clearly, t_a is an isometry.

Theorem 3.4.7. *Let $f \in \text{Iso}(\mathbb{R}^n)$. Then $f = f_A^b$ for some orthogonal real matrix A and $b \in \mathbb{R}^n$.*

Proof. Let $f : \mathbb{R}^n \to \mathbb{R}^n$ be an isometry. Let $b = f(0)$. Define the map $g : \mathbb{R}^n \to \mathbb{R}^n$ by $g = t_{-b} \circ f$. Note that g is isometry and $g(0) = 0$. By Proposition 3.4.6 there is an orthogonal matrix A such that

$$g(x) = Ax^t.$$

Therefore

$$f(x) = Ax^t + b. \qquad \square$$

Let $O(n, \mathbb{R})$ denote the set of all $n{\times}n$ real orthogonal matrices. Then $O(n, \mathbb{R})$ is a group under matrix multiplication. Define the binary operation on $O(n, \mathbb{R}) \times \mathbb{R}^n$ as follows:

$$(A_1, b_1)(A_2, b_2) = (A_1 A_2, b_1 + A_1 b_2).$$

Then $O(n, \mathbb{R}) \times \mathbb{R}^n$ is a group under this operation. We denote this group by $O(n, \mathbb{R}) \ltimes \mathbb{R}^n$. We can check that the map $\psi : O(n, \mathbb{R}) \ltimes \mathbb{R}^n \to \text{Iso}(\mathbb{R}^n)$ defined by

$$\psi((A, b)) = f_A^b$$

is a group isomorphism.

Now we find the isometries of $(\mathbb{S}^n, d_{\mathbb{S}^n})$. Let $x, y \in \mathbb{S}^n$. Then

$$\langle x, y \rangle = 1 - \frac{\|x - y\|^2}{2}.$$

This shows that $f : (\mathbb{S}^n, d_{\mathbb{S}^n}) \to (\mathbb{S}^n, d_{\mathbb{S}^n})$ is an isometry if and only if $f : (\mathbb{S}^n, d_{\mathbb{R}^{n+1}}) \to (\mathbb{S}^n, d_{\mathbb{R}^{n+1}})$ is an isometry. Therefore if f is an isometry of $(\mathbb{S}^n, d_{\mathbb{S}^n})$, then $f(0) = 0$.

Now suppose that $g : \mathbb{R}^{n+1} \to \mathbb{R}^{n+1}$ is an isometry such that $g(0) = 0$. Let $x, y \in \mathbb{S}^n$. Then

$$\begin{aligned}
\langle x, y \rangle &= 1 - \frac{\|x - y\|^2}{2} \\
&= 1 - \frac{\|g(x) - g(y)\|^2}{2} \\
&= \langle g(x), g(y) \rangle.
\end{aligned}$$

This implies that

$$d_{\mathbb{S}^n}(x, y) = d_{\mathbb{S}^n}(g(x), g(y)).$$

Therefore the restriction of g on $\mathbb{S}^n$ is an isometry of $(\mathbb{S}^n, d_{\mathbb{S}^n})$. Now suppose that $h : (\mathbb{S}^n, d_{\mathbb{S}^n}) \to (\mathbb{S}^n, d_{\mathbb{S}^n})$ is an isometry. Let $z_1, z_2 \in \mathbb{S}^n$. Then

$$d_{\mathbb{S}^n}(x,y) = d_{\mathbb{S}^n}(h(x),h(y)).$$

Therefore

$$\langle x,y \rangle = \langle h(x), h(y) \rangle.$$

Note that each nonzero element x of $\mathbb{R}^{n+1}$ can be uniquely written as $x = \|x\|\frac{x}{\|x\|}$. Definethea map $\bar{h} : \mathbb{R}^{n+1} \to \mathbb{R}^{n+1}$ by

$$\bar{h}(x) = \begin{cases} \|x\|h(\frac{x}{\|x\|}) & \text{if } x \neq 0, \\ 0 & \text{if } x = 0. \end{cases}$$

Let $x, y \in \mathbb{R}^{n+1}$ be such that $x \neq 0$ and $y \neq 0$. Note that

$$\begin{aligned} \langle \bar{h}(x), \bar{h}(y) \rangle &= \left\langle \|x\|h\left(\frac{x}{\|x\|}\right), \|y\|h\left(\frac{y}{\|y\|}\right) \right\rangle \\ &= \|x\|\|y\|\left\langle h\left(\frac{x}{\|x\|}\right), h\left(\frac{y}{\|y\|}\right) \right\rangle \\ &= \|x\|\|y\|\left\langle \frac{x}{\|x\|}, \frac{y}{\|y\|} \right\rangle \\ &= \langle x, y \rangle. \end{aligned} \qquad (3.16)$$

If $x = 0$ or $y = 0$, then equation (3.16) is obviously true. Now we claim that $\bar{h}$ is surjective. Let $y \in \mathbb{R}^{n+1}$. We have to find $x \in \mathbb{R}^{n+1}$ such that $\bar{h}(x) = y$. If $y = 0$, then $x = 0$. Suppose that $y \neq 0$. Note that if such x exists, then $x \neq 0$. This implies that for such x, we have $\|x\|h(\frac{x}{\|x\|}) = y$. Since $h(\frac{x}{\|x\|}) \in \mathbb{S}^n$, $\|x\| = \|y\|$. This shows that we have to find $x \in R^{n+1}$ such that $h(\frac{x}{\|x\|}) = \frac{y}{\|y\|}$. This is possible as h is an isometry.

Therefore, by Proposition 3.4.6, $\bar{h}$ is an isometry such that $\bar{h}(0) = 0$. Hence the isometries of $(\mathbb{S}^n, d_{\mathbb{S}^n})$ are precisely the restrictions of the isometries of $\mathbb{R}^{n+1}$ fixing the origin. Thus the group $\text{Iso}((\mathbb{S}^n, d_{\mathbb{S}^n}))$ is isomorphic to the group $O(n+1, \mathbb{R})$.

Consider the metric space $(\mathbb{R}^n, d_p)$ $(1 \leq p \leq \infty)$ of Examples 2.1.3 and 2.1.4. Note that the metric d_p is induced by the norm. We have obtained the isometries of $(\mathbb{R}^n, d_p)$ for $p = 2$ in Theorem 3.4.7. We have observed that the linear transformation of $\mathbb{R}^n$ that is an isometry is the map $f : \mathbb{R}^n \to \mathbb{R}^n$ given by $f(x) = Ax^t$, where A is an orthogonal matrix. Now we will obtain the isometries of $(\mathbb{R}^n, d_p)$ $(1 \leq p \leq \infty)$ that are linear transformations of $\mathbb{R}^n$. We will call such an isometry a linear isometry. If a linear isometry exists between the normed spaces V and W, then V and W are called linearly isometric spaces.

Let $S \subseteq \mathbb{R}^n$. A point $x \in S$ is called an extreme point of S if x cannot be written as $aa + (1-a)b$ with $a, b \in S$, $a \neq b$, and $0 < a < 1$. Let $\{e_1, \ldots, e_n\}$ be the standard basis of $\mathbb{R}^n$. Let E_p denote the set of the extreme points of the unit ball $B_{d_p}(0,1)$ in $(\mathbb{R}^n, d_p)$. Then we can check that

$$E_p = \begin{cases} \{\pm e_i \mid 1 \le i \le n\} & \text{if } p = 1, \\ \{x \in \mathbb{R}^n \mid \|x\|_p = 1\} & \text{if } 1 < p < \infty, \\ \{(\pm 1, \ldots, \pm 1) \mid 1 \le i \le n\} & \text{if } p = \infty. \end{cases}$$

Therefore

$$|E_p| = \begin{cases} 2n & \text{if } p = 1, \\ 2^n & \text{if } p = \infty, \end{cases}$$

and E_p is an uncountable set if $1 < p < \infty$.

Proposition 3.4.8. *Let V and W be linearly isometric normed spaces. Then the sets of extreme points of the unit balls $B_V(0,1)$ and $B_W(0,1)$ in V and W, respectively, are in a bijective correspondence.*

Proof. Let $f : V \to W$ be a linear isometry. Let $x \in B_V(0,1)$ be such that x is not an extreme point of $B_V(0,1)$. Then there exist $a, b \in B_V(0,1)$ with $a \ne b$ and $0 < \alpha < 1$ such that

$$x = \alpha a + (1 - \alpha)b.$$

This implies that

$$f(x) = \alpha f(a) + (1 - \alpha)f(b).$$

Note that $f(a), f(b) \in B_W(0,1)$ and $f(a) \ne f(b)$. This shows that $f(x)$ is not an extreme point of $B_W(0,1)$.

We can similarly show that if $y \in B_W(0,1)$ is not an extreme point of $B_W(0,1)$ with $f(x) = y$, then x in not an extreme point of $B_V(0,1)$. This shows that the set of extreme points of $B_V(0,1)$ is in a bijective correspondence with the set of extreme points of $B_W(0,1)$. □

Using Proposition 3.4.8, we can observe that the metric spaces $(\mathbb{R}^n, d_1)$ and $(\mathbb{R}^n, d_\infty)$ cannot be linearly isometric to $(\mathbb{R}^n, d_p)$ $(1 < p < \infty)$. Similarly, $(\mathbb{R}^n, d_\infty)$ cannot be linearly isometric to $(\mathbb{R}^n, d_1)$ if $n \ge 3$, since $2^n > 2n$ for $n \ge 3$. For $n = 1$, the metrics d_1 and d_∞ coincide, and we can check that the map $f : (\mathbb{R}^2, d_1) \to (\mathbb{R}^2, d_\infty)$ defined by $f(x,y) = (x - y, x + y)$ is a linear isometry.

Theorem 3.4.9 (Hemasinha and Weaver). *Let $f : (\mathbb{R}^n, d_1) \to (\mathbb{R}^n, d_1)$ be a linear isometry. Then $f(e_i) = \pm e_j$ for $1 \le i,j \le n$.*

Proof. Since f is bijective linear transformation, it sends a basis of $\mathbb{R}^n$ to a basis of $\mathbb{R}^n$. Since the set of extreme points of the unit ball $B(0,1)$ in $(\mathbb{R}^n, d_1)$ is

$$\{\pm e_i \mid 1 \le i \le n\},$$

by Proposition 3.4.8, $f(e_i) = \pm e_j$. □

Theorem 3.4.10 (Hemasinha and Weaver). *Let $f : (\mathbb{R}^n, d_\infty) \to (\mathbb{R}^n, d_\infty)$ be a linear isometry. Then $f(e_i) = \pm e_j$ for $1 \le i, j \le n$.*

Proof. Let A be the matrix of f with respect to the standard basis $\{e_1, \ldots, e_n\}$ of $\mathbb{R}^n$. Consider the set E_∞ of the extreme points of the unit ball $B(0, 1)$ in $(\mathbb{R}^n, d_\infty)$. Let $x \in E_\infty$. By Proposition 3.4.8, $Ax^t \in E_\infty$. Therefore, for all $1 \le j \le n$, we have

$$e_j Ax^t = \pm 1. \tag{3.17}$$

Note that for a fixed i, $e_i A$ denotes the ith row of the matrix A. Let

$$e_i A = (a_1, \ldots, a_n).$$

Suppose that

$$a_{i_1}, a_{i_2}, \ldots, a_{i_k} \le 0 \quad \text{and}$$
$$a_{i_{k+1}}, \ldots, a_{i_n} > 0,$$

where $a_{i_j} \in \{a_1, \ldots, a_n\}$. Taking $x = (1, 1, \ldots, 1)$ in equation (3.17), we get

$$e_i Ax^t = a_{i_1} + a_{i_2} + \cdots + a_{i_k} + a_{i_{k+1}} + \cdots + a_{i_n}$$
$$= \pm 1. \tag{3.18}$$

Now take $x = (\delta_1, \ldots, \delta_n)$ in equation (3.17), where

$$\delta_j = \begin{cases} -1 & \text{if } j = i_1, \ldots, i_k, \\ 1 & \text{if } j = i_{k+1}, \ldots, i_n. \end{cases}$$

Then we have

$$e_i Ax^t = -a_{i_1} - a_{i_2} + \cdots - a_{i_k} + a_{i_{k+1}} + \cdots + a_{i_n}$$
$$= \pm 1. \tag{3.19}$$

If the right sides of equations (3.18) and (3.19) are of equal sign, then subtracting (3.19) from (3.18), we get

$$a_{i_1} + a_{i_2} + \cdots + a_{i_k} = 0.$$

This implies that

$$a_{i_1} = a_{i_2} = \cdots = a_{i_k} = 0.$$

If the right sides of equations (3.18) and (3.19) are of opposite sign, then adding (3.18) and (3.19), we get

$$a_{i_{k+1}} + \cdots + a_{i_n} = 0.$$

This implies that

$$a_{i_{k+1}} = \cdots = a_{i_n} = 0.$$

This is only possible when

$$a_1, a_2, \ldots, a_n \leq 0.$$

Therefore, in both cases, we see that all nonzero components of $(a_1, \ldots, a_n)$ are of the same sign. Suppose that they are nonnegative and a_m is the first positive component. Take $x = (u_1, u_2, \ldots, u_n)$ in equation (3.17), where

$$u_j = \begin{cases} 1 & \text{if } 1 \leq j \leq m, \\ -1 & \text{if } m < j \leq n. \end{cases}$$

Then we have

$$e_i A x^t = a_m - a_{m+1} - \cdots - a_{i_n}$$
$$= \pm 1. \tag{3.20}$$

Now taking $x = (1, 1, \ldots, 1)$ in equation (3.17), we get

$$e_i A x^t = a_m + a_{m+1} + \cdots + a_{i_n}$$
$$= 1. \tag{3.21}$$

If the right sides of equations (3.20) and (3.21) are of opposite sign, then adding (3.20) and (3.21), we get $a_m = 0$, a contradiction. Therefore the right side of (3.20) is positive. Now subtracting (3.20) from (3.21), we get

$$a_{m+1} + \cdots + a_{i_n} = 0.$$

This implies that

$$a_{m+1} = \cdots = a_{i_n} = 0.$$

Therefore $a_m = 1$, and this is the only nonzero entry of the ith row $(a_1, \ldots, a_n)$ of A. Hence $f(e_i) = e_j$ for some j.

We can similarly show that $f(e_i) = -e_j$ when all the components of $(a_1, \ldots, a_n)$ are nonpositive. □

Now we will find the linear isometries of $(\mathbb{R}^n, d_p)$ $(1 < p < \infty)$. A linear isometry $f : \mathbb{R}^n \to \mathbb{R}^n$ is called nonnegative if all the components of $f(x)$ are nonnegative whenever all the components of x are nonnegative.

Proposition 3.4.11. *Let $f : (\mathbb{R}^n, d_p) \to (\mathbb{R}^n, d_p)$ $(1 < p < \infty)$ be a nonnegative linear isometry. Then $f(e_i) = e_j$.*

Proof. Let $A = (a_{ij})$ be the matrix of f with respect to the standard basis of $\mathbb{R}^n$. Then $a_{ij} \geq 0$ for all $1 \leq i, j \leq n$. Let $x = (x_1, \ldots, x_n)$ be such that $x_j \geq 0$ for all $1 \leq j \leq n$. Then

$$\|f(x)\|_p^p = \sum_{i=1}^{n} \left(\sum_{j=1}^{n} a_{ij} x_j \right)^p. \tag{3.22}$$

Since f is linear and isometry,

$$\begin{aligned} \|x\|_p &= \|x - 0\|_p \\ &= \|f(x) - f(0)\| \\ &= \|f(x)\|_p. \end{aligned}$$

This implies that

$$\sum_{i=1}^{n} x_i^p = \sum_{i=1}^{n} \left(\sum_{j=1}^{n} a_{ij} x_j \right)^p. \tag{3.23}$$

For $x = e_k$ $(1 \leq k \leq n)$ in equation (3.23), we have

$$1 = \sum_{i=1}^{n} a_{ik}. \tag{3.24}$$

We can consider (3.23) as the value of a differentiable function from $\mathbb{R}^n$ to $\mathbb{R}$. Now considering the partial derivative of (3.23) with respect to x_k, we have

$$x_k^{p-1} = \sum_{i=1}^{n} \left(\sum_{j=1}^{n} a_{ij} x_j \right)^{p-1} a_{ik}. \tag{3.25}$$

Putting $x_j = 1$ for $j \neq k$ and $x_k = 0$ in equation (3.25), we have

$$\sum_{i=1}^{n} \left(\sum_{j=1, j \neq k}^{n} a_{ij} \right)^{p-1} a_{ik} = 0. \tag{3.26}$$

Since $a_{ij} \geq 0$, we have

$$\left(\sum_{j=1, j \neq k}^{n} a_{ij} \right)^{p-1} a_{ik} = 0. \tag{3.27}$$

This implies that

$$\text{either} \quad a_{ik} = 0 \quad \text{or} \quad \sum_{j=1, j \neq k}^{n} a_{ij} = 0,$$

where $1 \leq i \leq n$. Since A is invertible (being the matrix of an isomorphism), there is no k such that $a_{ik} = 0$ for all i. Suppose that there is a kth column such that $a_{ik} \neq 0$. Further, suppose that $a_{rk} \neq 0$ and $a_{sk} \neq 0$, where $r \neq s$. Then from equation (3.27) we have

$$\sum_{j=1, j \neq k}^{n} a_{rj} = 0$$

and

$$\sum_{j=1, j \neq k}^{n} a_{sj} = 0.$$

This implies that $a_{rj} = 0$ and $a_{sj} = 0$ for $j \neq k$. This shows that the rth row of A is a multiple of the sth row. This is a contradiction, since A is invertible. This implies that each column and each row of the matrix A contains exactly one nonzero entry. By equation (3.24) we see that A is the matrix obtained by permuting the row of the identity matrix. Hence

$$f(e_i) = e_j. \qquad \qquad \square$$

Proposition 3.4.12. *Let $f : (\mathbb{R}^n, d_p) \to (\mathbb{R}^n, d_p)$ ($1 < p < \infty$ and $p \neq 2$) be a linear isometry. If all the components of $x \in \mathbb{R}^n$ are nonzero, then all the components of $f(x)$ are nonzero.*

Proof. On the contrary, suppose that there is $x = (x_1, \ldots, x_n) \in \mathbb{R}^n$ with $x_i \neq 0$ for all i but $f(x) = (y_1 \ldots, y_n)$ has some component $y_k = 0$. Let $z = (z_1, \ldots, z_n) \in \mathbb{R}^n$ be such that $f(z) = e_k$, where $e_k = (0, \ldots, 0, 1, 0, \ldots, 0)$. Note that for any $\alpha \in \mathbb{R}$, we have

$$\|f(x + \alpha z)\|_p^p = \|f(x) + \alpha f(z)\|_p^p$$
$$= \|f(x)\|_p^p + \|\alpha f(z)\|_p^p.$$

Since f is isometry,

$$\|x + \alpha z\|_p^p = \|x\|_p^p + \|\alpha z\|_p^p.$$

This implies that

$$\sum_{i=1}^{n} |x_i + \alpha z_i|^p = \sum_{i=1}^{n} |x_i|^p + \sum_{i=1}^{n} |\alpha z_i|^p. \qquad (3.28)$$

For given x and z, we can view equation (3.28) as the value of a twice differentiable map in α from $\mathbb{R}$ to $\mathbb{R}$. Differentiating (3.28) twice with respect to α, we get

$$\sum_{i=1}^{n} p(p-1)|x_i + \alpha z_i|^{p-2} z_i^2 = |\alpha|^{p-2} p(p-1) \sum_{i=1}^{n} |z_i|^p.$$

This implies that

$$\sum_{i=1}^{n} |x_i + \alpha z_i|^{p-2} z_i^2 = |\alpha|^{p-2} \sum_{i=1}^{n} |z_i|^p. \qquad (3.29)$$

First, suppose that $p > 2$. Taking $\alpha = 0$ in equation (3.29), we get

$$\sum_{i=1}^{n} |x_i|^{p-2} z_i^2 = 0.$$

Since $|x_i| > 0$ for all i, $z_i = 0$ for all i. This is a contradiction. Therefore $p < 2$. Now equation (3.29) can be written as

$$|\alpha|^{2-p} \sum_{i=1}^{n} |x_i + \alpha z_i|^{p-2} z_i^2 = \sum_{i=1}^{n} |z_i|^p. \qquad (3.30)$$

Taking $\alpha = 0$ in equation (3.30), we get

$$\sum_{i=1}^{n} |z_i|^p = 0.$$

This implies that $z_i = 0$ for all i, a contradiction. Therefore all the components of $f(x)$ are nonzero. □

Proposition 3.4.13. *Let $f : (\mathbb{R}^n, d_p) \to (\mathbb{R}^n, d_p)$ $(1 < p < \infty$ and $p \neq 2)$ be a linear isometry. Let $u = (1, 1, \dots, 1)$. If all the components of $f(u)$ are positive, then f is nonnegative.*

Proof. We will prove that if all the components of $f(u)$ are positive, then all the components of $f(e_i)$ are nonnegative for all i. This will show that f is nonnegative, because for $x = (x_1, \dots, x_n)$,

$$f(x) = x_1 f(e_1) + \cdots + x_n f(e_n).$$

On the contrary, suppose that there is j such that $f(e_j) = (y_1, \dots, y_n)$ has some component $y_k < 0$. Let $f(u) = (a_1, \dots, a_n)$, and let $\alpha = \frac{-a_k}{y_k}$. Then $\alpha > 0$. Now the kth component of

$f(u + ae_j) = f(u) + af(e_j)$ is zero, but all the components of $u + ae_j$ are positive. This is a contradiction to Proposition 3.4.12. Hence all the components of $f(e_i)$ are nonnegative for all i. □

Theorem 3.4.14 (Hemasinha and Weaver). *Let $f : (\mathbb{R}^n, d_p) \rightarrow (\mathbb{R}^n, d_p)$ $(1 < p < \infty$ and $p \neq 2)$ be a linear isometry. Then $f(e_i) = \pm e_j$ for $1 \leq i, j \leq n$.*

Proof. Let $f(u) = (a_1, \dots, a_n)$, where $u = (1, 1, \dots, 1)$. By Proposition 3.4.12, $a_i \neq 0$ for all i. Define

$$\epsilon_i = \begin{cases} 1 & \text{if } a_i > 0, \\ -1 & \text{if } a_i < 0. \end{cases}$$

Let us consider the linear maps $g : \mathbb{R}^n \rightarrow \mathbb{R}^n$ defined by

$$g(e_i) = \epsilon_i e_i, \quad 1 \leq i \leq n.$$

Let $h = g \circ f$. Note that h is a linear isometry and

$$h(u) = (\epsilon_1 a_1, \dots, \epsilon_n a_n).$$

Observe that the components of $h(u)$ are nonnegative. By Proposition 3.4.13, $h(e_i) = e_j$. Note that $g^{-1} = g$. This implies that $f = g \circ h$. Hence $f(e_i) = \pm e_j$. □

3.5 Finite metric spaces

A metric space (X, d) is called finite if the set X contains finitely many points. As an example of a finite metric space, take a finite subset of some given metric space and consider it as a metric subspace of X. We can naturally ask that given a finite metric space X, can we embed X isometrically in some Euclidean space $(\mathbb{R}^n, d_2)$? This is not always possible as the following example shows.

Example 3.5.1. Let $X = \{p_0, p_1, p_2, p_3\}$. Define the metric d on X by $d(p_0, p_1) = d(p_0, p_2) = d(p_1, p_2) = 2$, $d(p_0, p_3) = d(p_1, p_3) = 1$, and $d(p_2, p_3) = \frac{3}{2}$. We cannot embed X isometrically in any Euclidean space. Indeed, if X is embedded isometrically in some Euclidean space, then the points p_0, p_1, and p_2 will be the vertices of an equilateral triangle of length 2. Also, p_3 is the midpoint of p_0 and p_1. We can check that the distance of p_2 and p_3 cannot be $\frac{3}{2}$.

We can always embed a finite metric space X containing n elements in $(\mathbb{R}^n, d_\infty)$, as the following example shows.

Example 3.5.2. Let (X, d) be a metric space, where $X = \{p_0, p_1, \dots, p_n\}$. Consider the $((n+1) \times (n+1))$ matrix $A = (a_{ij})$, where $a_{ij} = d(p_i, p_j)$. Let $A_i = (a_{i0}, a_{i1}, \dots, a_{in})$. Define the map $\phi : (X, d) \rightarrow (\mathbb{R}^{n+1}, d_\infty)$ by $\phi(p_i) = A_i$.

By the triangle inequality $a_{ik} + a_{kj} \geq a_{ij}$ we can observe that

$$d_\infty(A_i, A_j) = \|A_i - A_j\|_\infty \leq a_{ij} \quad \text{for all } i, j.$$

Note that $|a_{ij} - a_{jj}| = a_{ij}$. This implies that the above inequality is indeed an equality, that is, $d_\infty(A_i, A_j) = a_{ij}$.

Now

$$d_\infty(\phi(p_i), \phi(p_j)) = \|A_i - A_j\|_\infty$$
$$= a_{ij}$$
$$= d(p_i, p_j).$$

Now we move to the question of embedding a finite metric space isometrically into the Euclidean space. Let $X = \{p_0, p_1, \ldots, p_n\}$. Consider the matrices $M = (m_{ij})$ and $G = (g_{ij})$, where $m_{ij} = d(p_i, p_j)^2$ and $g_{ij} = \frac{1}{2}(m_{0i} + m_{0j} - m_{ij})$. The matrix G is called the Gramian matrix of X. Note that the Gramian matrix of X is an $(n+1) \times (n+1)$ matrix whose first row and first column are zero. This can be represented as follows:

$$G = \left(\begin{array}{c|c} 0 & \mathbf{0} \\ \hline \mathbf{0} & G_1 \end{array} \right),$$

where G_1 is an $n \times n$ matrix. Note that G is positive semidefinite if and only if G_1 is positive semidefinite.

Theorem 3.5.3 (Schoenberg). *A finite matric space (X, d), where $X = \{p_0, p_1, \ldots, p_n\}$, is isometrically embedded in $(\mathbb{R}^n, d_2)$ if and only if its Gramian matrix G is positive semidefinite and has rank at most n.*

Proof. Suppose that X is isometrically embedded in $(\mathbb{R}^n, d_2)$. Suppose $\phi : X \to \mathbb{R}^n$ is an isometric embedding. Without loss of generality, we can assume that $\phi(p_0) = 0 \in \mathbb{R}^n$. Let

$$\phi(p_i) = (a_{i1}, \ldots, a_{in}), \quad 1 \leq i \leq n.$$

Viewing $x = (x_1, \ldots, x_n) \in \mathbb{R}^n$ as $1 \times n$ matrix, let $Q(x) = xG_1x^t$ be the quadratic form associated with G_1, where G_1 is the $n \times n$ submatrix mentioned above. Let

$$P = x_1\phi(p_1) + \cdots + x_n\phi(p_n) \in \mathbb{R}^n.$$

Then

$$\|P\|_2 = \sum_{k=1}^{n} (x_1a_{1k} + \cdots + x_na_{nk})^2$$
$$= \sum_{i=1}^{n} x_i^2 \sum_{k=1}^{n} a_{ik}^2 + 2\sum_{i<j} x_ix_j \sum_{k=1}^{n} a_{ik}a_{jk}.$$

Note that

$$\sum_{k=1}^{n} a_{ik}^2 = \|\phi(p_i)\|_2^2$$
$$= \|\phi(p_i) - \phi(p_0)\|_2^2$$
$$= d(p_i, p_0)^2$$
$$= m_{0i}$$
$$= g_{ii}$$

and

$$2\sum_{k=1}^{n} a_{ik}a_{jk} = \sum_{k=1}^{n} a_{ik}^2 + \sum_{k=1}^{n} a_{jk}^2 - \sum_{k=1}^{n}(a_{ik} - a_{jk})^2$$
$$= \|\phi(p_i)\|_2^2 + \|\phi(p_j)\|_2^2 - \|\phi(p_i) - \phi(p_j)\|_2^2$$
$$= d(p_i, p_0)^2 + d(p_j, p_0)^2 - d(p_i, p_j)^2$$
$$= m_{0i} + m_{0j} - m_{ij}$$
$$= 2g_{ij}.$$

Then

$$\|P\|_2^2 = \sum_{i=1}^{n} m_{0i}x_i^2 + \sum_{i<j} g_{ij}x_i x_j$$
$$= xGx^t$$
$$= Q(x).$$

This implies that $Q(x) \geq 0$. Therefore G_1 is positive semidefinite.

Suppose $Q(x) = 0$. Then $P = 0$. This implies that

$$x_1\phi(p_1) + \cdots + x_n\phi(p_n) = 0.$$

Note that $\phi(i)$ is the ith row of G_1. This shows that the rank of G_1 is at most n.

Conversely, suppose that G is positive semidefinite and the rank of G is $r \leq n$. Let G_1 be the $n \times n$ submatrix mentioned above. Then G_1 is positive semidefinite. Since G_1 is a symmetric matrix, it is diagonalizable. Then there is an orthogonal matrix U such that $G_1 = UDU^t$, where D is the diagonal matrix $\text{diag}(d_1, \ldots, d_r, \ldots, 0, \ldots, 0)$ with $d_i \neq 0$. Since G_1 is positive semidefinite, $d_i > 0$ for $1 \leq i \leq r$. Let

$$\sqrt{D} = \text{diag}(\sqrt{d_1}, \ldots, \sqrt{d_r}, \ldots, 0, \ldots, 0)$$

and $P = \sqrt{D}U^t$. Then

$$P^t P = U \sqrt{D} \sqrt{D} U^t = G_1.$$

Define $\phi : X \to \mathbb{R}^n$ by

$$\phi(p_i) = i\text{th column of } P(1 \le i \le n)$$

and $\phi(p_0) = 0$. Then

$$P = [\phi(p_1), \ldots, \phi(p_n)].$$

The ijth entry of $P^t P$ is $\langle \phi(p_i), \phi(p_j) \rangle$. Therefore $g_{ij} = \langle \phi(p_i), \phi(p_j) \rangle$. Now

$$\begin{aligned}
\|\phi(p_i)\|_2^2 &= \langle \phi(p_i), \phi(p_i) \rangle \\
&= g_{ii} = m_{0i} \\
&= d(a_i, a_0)^2.
\end{aligned}$$

This shows that

$$\|\phi(p_i) - \phi(p_0)\|_2 = d(p_i, p_0).$$

Again,

$$\begin{aligned}
\|\phi(p_i) - \phi(p_j)\|_2^2 &= \|\phi(p_i)\|_2 + \|\phi(p_j)\|_2 - 2\langle \phi(p_i), \phi(p_j) \rangle \\
&= m_{0i} + m_{0j} - (m_{0i} + m_{0j} - m_{ij}) \\
&= m_{ij} \\
&= d(p_i, p_j)^2.
\end{aligned}$$

This shows that $\|\phi(p_i) - \phi(p_j)\|_2 = d(p_i, p_j)$. Hence ϕ is an isometric embedding of X into $\mathbb{R}^n$. $\qquad\square$

Given the metric space $X = \{p_0, p_1, \ldots, p_n\}$, the determinant

$$D(p_0, p_1, \ldots, p_n) = \begin{vmatrix} 0 & 1 & 1 & \cdots & 1 \\ 1 & 0 & m_{01} & \cdots & m_{0n} \\ 1 & m_{10} & 0 & \cdots & m_{1n} \\ \vdots & \vdots & \vdots & \ddots & \vdots \\ 1 & m_{n0} & m_{n1} & \cdots & 0 \end{vmatrix}$$

is called the Cayley–Menger determinant. We can prove that

$$\det G = \frac{(-1)^{n+1}}{2^n} D(p_0, p_1, \ldots, p_n), \tag{3.31}$$

where G is the Gramian matrix of $X = \{p_0, p_1, \ldots, p_n\}$. The following is a result from linear algebra, whose proof can be found in [4].

Theorem 3.5.4. *Let A be an $n \times n$ real symmetric matrix of rank r. Then A is positive semidefinite if and only if the first r leading principal minors of A are positive and $\det A \geq 0$.*

By Theorems 3.5.3 and 3.5.4 and equation (3.31) we obtain the following:

Theorem 3.5.5. *A finite metric space X is isometrically embedded in the Euclidean $\mathbb{R}^n$ but not in $\mathbb{R}^{n-1}$ if and only if there are $p_0, p_1, \ldots, p_n \in X$ such that*
(i) $(-1)^{j+1} D(p_0, p_1, \ldots, p_j) > 0$ *for* $1 \leq j \leq n$ *and*
(ii) $D(p_0, p_1, \ldots, p_n, x) = D(p_0, p_1, \ldots, p_n, y) = D(p_0, p_1, \ldots, p_n, x, y) = 0$ *for all* $x, y \in X$.

Exercises

3.1. First of all, complete whatever is left for you as exercises.
3.2. Show that the function $f : \mathbb{R} \to \mathbb{R}$ be defined by

$$f(x) = \begin{cases} 0, & x \leq 0, \\ x, & x \geq 0, \end{cases}$$

is continuous.
3.3. Show that the maps $f : \mathbb{R}^2 \to \mathbb{R}$ and $g : \mathbb{R}^2 \to \mathbb{R}$ defined by $f(x, y) = x + y$ and $g(x, y) = xy$ are continuous.
3.4. Show that the map $f : \mathbb{R}^n \times \mathbb{R}^n \to \mathbb{R}$ defined by $f(x, y) = \langle x, y \rangle$ is continuous.
3.5. If f and g are continuous maps from a metric space X to $\mathbb{R}$, then show that the map $(f, g) : X \to \mathbb{R}$ defined by $(f, g)(x) = (f(x), g(x))$ is continuous.
3.6. If f and g are continuous maps from a metric space X to a metric space Y, then show that the set

$$\{x \in X \mid f(x) = g(x)\}$$

is a closed set in X.
3.7. Let f and g be continuous maps from a metric space X to a metric space Y. If they are equal on some dense set in X, then show that $f = g$.
3.8. Show that $\mathbb{S}^n \setminus \{e_{n+1}\}$ is homeomorphic to $\mathbb{R}^n$, where $e_{n+1} = (0, \ldots, 0, 1)$.
3.9. Show that the closed ball $D(x, r)$ in $\mathbb{R}^n$ is homeomorphic to $[0, 1]^n$.
3.10. Check whether the map $f : (\frac{-\pi}{2}, \frac{\pi}{2}) \to \mathbb{R}$ defined by $f(x) = \tan x$ is uniformly continuous or not.
3.11. Let V be an infinite-dimensional normed space. Show that there is a discontinuous functional on V.

3.12. Show that the map $f : X \to Y$ is continuous if and only if

$$d_X(x, A) = 0 \Rightarrow d_Y(f(x), f(A)) = 0$$

for all $A \subseteq X$ and $x \in X$.

3.13. Let X be a metric space, and let

$$\{f_\alpha : X \to [a, b] \mid \alpha \in \mathcal{A}\}$$

be a family of continuous maps. Check whether the maps

$$\mu(x) = \mathrm{lub}\{f_\alpha(x) \mid \alpha \in \mathcal{A}\}$$

and

$$\kappa(x) = \mathrm{glb}\{f_\alpha(x) \mid \alpha \in \mathcal{A}\}$$

are continuous or not.

3.14. Give an example of a finite metric space X that cannot be isometrically embedded in $\mathbb{Z}^n$. Can you think of some conditions such that X can be isometrically embedded in $\mathbb{Z}^n$?

3.15. Let A and B be two bounded polygons in the plane. Show that

$$\delta(A, B) = a(A) + a(B) - 2a(A \cap B)$$

is a metric on set of all bounded polygons in a plane, where $a(A)$ denotes the area of A. Also, show that it is not topologically equivalent to the Hausdorff distance.

3.16. Show that the metric δ defined in the Exercise 3.15. is topologically equivalent to the Hausdorff distance if we restrict δ to the set of convex bounded polygons in the plane.

4 Product and quotient metric spaces

In this chapter, we study the metrics defined on the product and quotient of metric spaces.

4.1 Product of metric spaces

Let $\{(X_\alpha, d_\alpha) \mid \alpha \in \Lambda\}$ be a family of metric spaces. We can ask whether the product $\prod_{\alpha \in \Lambda} X_\alpha$ is a metric space. It is obvious that we can define the discrete metric on $\prod_{\alpha \in \Lambda} X_\alpha$. We can further ask whether there is a metric on $\prod_{\alpha \in \Lambda} X_\alpha$ using the metrics on $X_\alpha, \alpha \in \Lambda$. We will answer this question by considering different possibilities on the indexing set Λ. We are indeed interested in the minimal possibility of open sets in $\prod_{\alpha \in \Lambda} X_\alpha$ such that the projection maps are continuous. We will discuss this after we define the topology.

Example 4.1.1. Let $(X_1, d_1), \ldots, (X_n, d_n)$ be n metric spaces. Let $x = (x_1, \ldots, x_n)$ and $y = (y_1, \ldots, y_n)$ be in $\prod_{i=1}^{n} X_i$. Define $d : \prod_{i=1}^{n} X_i \times \prod_{i=1}^{n} X_i \to \mathbb{R}$ by

$$d(x, y) = \left(\sum_{i=1}^{n} d_i(x_i, y_i)^2 \right)^{\frac{1}{2}}.$$

We can check that d is a metric on $\prod_{i=1}^{n} X_i$.

Example 4.1.2. Let $\| \cdot \|$ be a norm on $\mathbb{R}^n$. Let $(X_1, d_1), \ldots, (X_n, d_n)$ be n metric spaces. Let $x = (x_1, \ldots, x_n), y = (y_1, \ldots, y_n)$, and $z = (z_1, \ldots, z_n)$ be in $\prod_{i=1}^{n} X_i$. Define $d : \prod_{i=1}^{n} X_i \times \prod_{i=1}^{n} X_i \to \mathbb{R}$ by

$$d(x, y) = \|(d_1(x_1, y_1), \ldots, d_n(x_n, y_n))\|.$$

We can check that d is nonnegative and symmetric and that $d(x, y) = 0$ if and only if $x = y$. For the triangle inequality, let $x, y, z \in \prod_{i=1}^{n} X_i$. Then

$$\begin{aligned}
d(x, y) + d(y, z) &= \|(d_1(x_1, y_1), \ldots, d_n(x_n, y_n))\| \\
&\quad + \|(d_1(y_1, z_1), \ldots, d_n(y_n, z_n))\| \\
&\geq \|(d_1(x_1, y_1) + d_1(y_1, z_1), \ldots, d_n(x_n, y_n) + d_n(y_n, z_n))\| \\
&\geq \|(d_1(x_1, z_1), \ldots, d_n(x_n, z_n))\| \\
&= d(x, z).
\end{aligned}$$

Hence d is a metric on $\prod_{i=1}^{n} X_i$.

Later, we will prove that all the norms of a finite-dimensional vector space are Lipschitz equivalent. Therefore it is sufficient to deal with one metric on $\prod_{i=1}^{n} X_i$ if it is de-

https://doi.org/10.1515/9783111636085-004

fined with the help of a norm on $\mathbb{R}^n$ as given in Example 4.1.2 (as far as open sets or closed sets are concerned).

Let $\{(X_i, d_i) \mid 1 \le i \le n\}$ be a set of n metric spaces, where $n \in \mathbb{N}$. Consider the norm $\|\cdot\|_\infty$ on $\mathbb{R}^n$ as defined in Example 2.2.4. Define a metric d_∞ on $\prod_{i=1}^n X_i$ using the norm $\|\cdot\|_\infty$ as defined in Example 4.1.2. Then we can check that for $x = (x_1, \ldots, x_n) \in \prod_{i=1}^n X_i$ and $r > 0$, we have

$$B_{d_\infty}(x, r) = B_{d_1}(x_1, r) \times \cdots \times B_{d_n}(x_n, r).$$

Proposition 4.1.3. *Let $\{(X_i, d_i) \mid 1 \le i \le n\}$ be a set of n metric spaces, where $n \in \mathbb{N}$. Let V be an open set in $(\prod_{i=1}^n X_i, d_\infty)$, and let $x \in V$. Then there is an open set $\prod_{i=1}^n U_i$ in $(\prod_{i=1}^n X_i, d_\infty)$ such that*

$$x \in \prod_{i=1}^n U_i \subseteq V,$$

where U_i is open in (X_i, d_i).

Proof. We first show that if U_i is open in (X_i, d_i), then $\prod_{i=1}^n U_i$ is open in $(\prod_{i=1}^n X_i, d_\infty)$.

Let $z = (z_1, \ldots, z_n) \in \prod_{i=1}^n U_i$. Since each U_i is open in X_i, there is a positive real number r_i such that

$$B_{d_i}(z_i, r_i) \subseteq U_i.$$

Let $r = \min\{r_1, \ldots, r_n\}$. Then

$$B_{d_\infty}(z, r) = B_{d_1}(z_1, r) \times \cdots \times B_{d_n}(z_n, r) \subseteq \prod_{i=1}^n U_i.$$

Now let V be an open set in $\prod_{i=1}^n X_i$, and let $x = (x_1, \ldots, x_n) \in V$. Since V is open in $\prod_{i=1}^n X_i$, there is $r > 0$ such that

$$B_{d_\infty}(x, r) \subseteq V.$$

Note that $B_{d_\infty}(x, r) = B_{d_1}(x_1, r) \times \cdots \times B_{d_n}(x_n, r)$ and $U_i = B_{d_i}(x_i, r)$ is open in X_i. This completes the proof. $\square$

Corollary 4.1.4. *Let $\{(X_i, d_i) \mid 1 \le i \le n\}$ be a set of n metric spaces, where $n \in \mathbb{N}$. Then each open set in $(\prod_{i=1}^n X_i, d_\infty)$ is a union of open sets of the form $\prod_{i=1}^n U_i$, where U_i is open in (X_i, d_i).*

Corollary 4.1.5. *Let $\{(X_i, d_i) \mid 1 \le i \le n\}$ be a set of n metric spaces, where $n \in \mathbb{N}$. Then the projection maps $\pi_i : \prod_{i=1}^n X_i \to X_i$ defined by $\pi_i(x_1, \ldots, x_n) = x_i$ are continuous.*

Proof. Let U_i be open sets in X_i. Then

$$\pi_i^{-1}(U_i) = \prod_{j=1}^{n} V_j,$$

where

$$V_j = \begin{cases} U_i & \text{if } j = i, \\ X_j & \text{otherwise.} \end{cases} \qquad \square$$

Let us consider a countable family $\{(X_i, d_i) \mid i \in \mathbb{N}\}$ of metric spaces. The idea of giving a metric on $\prod_{i=1}^{\infty} X_i$ with the help of a norm as defined in Example 4.1.2 may not work in this case. It may yield a divergent series. For example, let $X_i = \mathbb{R}$, and let d_i be the discrete metric on X_i for all $i \in \mathbb{N}$. Consider the ℓ_1 norm $\|\cdot\|_1$. Define $d : \prod_{i=1}^{\infty} X_i \times \prod_{i=1}^{\infty} X_i \to \mathbb{R}$ by

$$d((x_n), (y_n)) = \sum_{i=1}^{\infty} d_i(x_n, y_n).$$

Let (a_n) and (b_n) be two elements in $\prod_{i=1}^{\infty} X_i$ such that $a_n \neq b_n$ for all $n \in \mathbb{N}$. Then

$$d((a_n), (b_n)) = \sum_{i=1}^{\infty} 1,$$

which is not a convergent series. For a collection $\{(X_i, d_i) \mid i \in \mathbb{N}\}$ of metric spaces, we define $d : \prod_{i=1}^{\infty} X_i \times \prod_{i=1}^{\infty} X_i \to \mathbb{R}$ by

$$d(x, y) = \sum_{i=1}^{\infty} \frac{1}{2^i} \frac{d_i(x_i, y_i)}{1 + d_i(x_i, y_i)}$$

for $x = (x_n)$ and $y = (y_n)$. We can check that d is a metric on the product $\prod_{i=1}^{\infty} X_i$. We can also define the metric d on $\prod_{i=1}^{\infty} X_i$ as follows:

$$d(x, y) = \sum_{i=1}^{\infty} a_i \frac{d_i(x_i, y_i)}{1 + d_i(x_i, y_i)},$$

where $\sum a_i$ is a convergent series of positive terms. If each d_i were a bounded metric of diameter at most 1, then we could have defined the metric d on $\prod_{i=1}^{\infty} X_i$ as follows:

$$d(x, y) = \sum_{i=1}^{\infty} \frac{d_i(x_i, y_i)}{2^i}. \tag{4.1}$$

Proposition 4.1.6. *Let $\{(X_i, d_i) \mid i \in \mathbb{N}\}$ be a collection of bounded metric spaces of diameter at most 1. Let d be the metric on the product $\prod_{i=1}^{\infty} X_i$ as defined in equation (4.1). Let V* be an open set in $(\prod_{i=1}^\infty X_i, d)$, and let $x \in V$. Then there are $n \in \mathbb{N}$ and an open set $\prod_{i=1}^\infty U_i$ in $(\prod_{i=1}^\infty X_i, d_\infty)$, where U_i is open in X_i for $1 \le i \le n$ and $U_i = X_i$ for all $i > n$ such that

$$x \in \prod_{i=1}^\infty U_i \subseteq V.$$

Proof. Let $n \in \mathbb{N}$. We first show that $\prod_{i=1}^\infty U_i$, where U_i is open in X_i and $U_i = X_i$ for all $i > n$, is open in $\prod_{i=1}^\infty X_i$. Let $z = (z_i) \in \prod_{i=1}^\infty U_i$. For each $1 \le i \le n$, there is $r_i > 0$ such that

$$B_{d_i}(z_i, r_i) \subseteq U_i.$$

Let

$$r = \min\left\{\frac{r_1}{2}, \frac{r_2}{2^2}, \dots, \frac{r_n}{2^n}\right\}.$$

Let $y = (y_i) \in B_d(z, r)$. Then for each $1 \le i \le n$, we have

$$\frac{d_i(z_i, y_i)}{2^i} \le d(z, y) < r \le \frac{r_i}{2^i}.$$

Therefore $d_i(z_i, y_i) < r_i$. This implies that

$$y = (y_i) \in B_{d_1}(z_1, r_1) \times \cdots \times B_{d_n}(z_n, r_n) \times X_{n+1} \times X_{n+2} \times \cdots$$
$$\subseteq \prod_{i=1}^\infty U_i.$$

Since $y \in B_d(z, r)$ is arbitrary, $B_d(z, r) \subseteq \prod_{i=1}^\infty U_i$. This shows that z is an interior point of $\prod_{i=1}^\infty U_i$. Since $z \in \prod_{i=1}^\infty U_i$ is arbitrary, $\prod_{i=1}^\infty U_i$ is open.

Now let V be an open set in $\prod_{i=1}^\infty X_i$, and let $x = (x_i) \in V$. Then there is $r > 0$ such that

$$B_d(x, r) \subseteq V.$$

Choose $n \in \mathbb{N}$ such that

$$\sum_{i=n+1}^\infty \frac{1}{2^i} < \frac{r}{2}.$$

Consider the open set

$$B = B_{d_1}\left(x_1, \frac{r}{2n}\right) \times \cdots \times B_{d_n}\left(x_n, \frac{r}{2n}\right) \times X_{n+1} \times X_{n+2} \times \cdots.$$

Note that $x \in B$. Let $y = (y_i) \in B$. Then $d_i(x_i, y_i) < \frac{r}{2n}$ for all $1 \le i \le n$. Then

$$d(x,y) = \sum_{i=1}^{\infty} \frac{d_i(x_i, y_i)}{2^i}$$

$$= \sum_{i=1}^{n} \frac{d_i(x_i, y_i)}{2^i} + \sum_{i=n+1}^{\infty} \frac{d_i(x_i, y_i)}{2^i}$$

$$< n\frac{r}{2n} + \frac{r}{2}$$

$$= r.$$

Therefore $y \in B_d(x, r)$. This shows that $B \subseteq B_d(x, r)$. Thus we have

$$x \in B \subseteq V. \qquad \square$$

We can show that an open set in $(\prod_{i=1}^{\infty} X_i, d)$ is the union of open sets $\prod_{i=1}^{\infty} U_i$, where U_i is open in X_i, and $U_i = X_i$ for all $i > n$. We can also show that each projection map is continuous. Note that

$$\prod_{i=1}^{n} \pi_i^{-1}(U_i) = \prod_{i=1}^{\infty} U_i,$$

where U_i is open in X_i, and $U_i = X_i$ for all $i > n$.

Let F_i be closed in X_i. Then

$$\prod_{i=1}^{\infty} F_i = \bigcap_{i=1}^{\infty} \pi_i^{-1}(F_i).$$

This shows that $\prod_{i=1}^{\infty} F_i$ is closed in $\prod_{i=1}^{\infty} X_i$.

Proposition 4.1.7. *Let X be a nonempty set, and let Z be the set of all sequences in X. Let $x = (x_n), y = (y_n) \in Z$. Define $\rho : Z \times Z \to \mathbb{R}$ by*

$$\rho(x, y) = \begin{cases} 0 & \text{if } x = y, \\ \frac{1}{\min\{i \in \mathbb{N} \mid x_i \neq y_i\}} & \text{otherwise.} \end{cases}$$

Then ρ is a metric on Z.

Proof. We will only show the triangle inequality. Let $x = (x_n), y = (y_n)$, and $z = (z_n)$ be in Z. If any two of x, y, and z are equal, then the triangle inequality holds. Suppose that $x \neq y, y \neq z$, and $z \neq x$. Let

$$l = \min\{i \in \mathbb{N} \mid x_i \neq y_i\}, \quad k = \min\{i \in \mathbb{N} \mid y_i \neq z_i\},$$
$$m = \min\{i \in \mathbb{N} \mid z_i \neq x_i\}.$$

Note that for $i \in \mathbb{N}$, if $x_i = y_i$ and $y_i = z_i$, then $x_i = z_i$. Therefore

$$m \geq \min\{l, k\}.$$

This implies that

$$\frac{1}{m} \leq \frac{1}{l} + \frac{1}{k}.$$

This proves the triangle inequality. □

Proposition 4.1.8. *Let (X, d') be the discrete metric space, and let Z be a set of sequences in X. Then the metric space (Z, ρ) is topologically equivalent to the metric space $(\prod_{i=1}^{\infty} X, d)$, where ρ and d are the metrics defined in equation* (4.1) *and* 4.1.7, *respectively.*

Proof. Note that if U is an open set in (Z, ρ) containing $x = (x_i) \in Z$, then $B(x, \frac{1}{n}) \subseteq U$ for some $n \in \mathbb{N}$. Let $y = (y_i) \in Z$. Then

$$y \in B_\rho\left(x, \frac{1}{n}\right) \Leftrightarrow d(x, y) < \frac{1}{n}$$

$$\Leftrightarrow \frac{1}{\min\{i \in \mathbb{N} \mid x_i \neq y_i\}} < \frac{1}{n}$$

$$\Leftrightarrow n < \min\{i \in \mathbb{N} \mid x_i \neq y_i\}$$

$$\Leftrightarrow x_i = y_i \text{ for all } 1 \leq i \leq n$$

$$\Leftrightarrow y \in \{x_1\} \times \cdots \times \{x_n\} \times X \times X \times \cdots.$$

This shows that

$$B_\rho\left(x, \frac{1}{n}\right) = \{x_1\} \times \cdots \times \{x_n\} \times X \times X \times \cdots.$$

By Proposition 4.1.6 and the discussion above we can observe that U is open in (Z, ρ) if and only if U is open in $(\prod_{i=1}^{\infty} X, d)$. □

Let us consider an arbitrary family $\{(X_\alpha, d_\alpha) \mid \alpha \in \Lambda\}$ of metric spaces. Let $\bar{d}_\alpha$ be the standard bounded metric on X_α corresponding to d_α. Let $x = (x_\alpha)$ and $y = (y_\alpha)$ be two elements in $\prod_{\alpha \in \Lambda} X_\alpha$. Define $d : \prod_{\alpha \in \Lambda} X_\alpha \times \prod_{\alpha \in \Lambda} X_\alpha \to \mathbb{R}$ by

$$d(x, y) = \sup\{\bar{d}_\alpha(x_\alpha, y_\alpha) \mid \alpha \in \Lambda\}.$$

We can check that d is a metric on $\prod_{\alpha \in \Lambda} X_\alpha$. This metric is called the uniform metric on the product $\prod_{\alpha \in \Lambda} X_\alpha$.

4.2 Metric-preserving maps

Let us revisit Example 4.1.1. Consider it for two metric spaces (X, d_X) and (Y, d_Y). The metric d on $X \times Y$ is defined by

$$d((x_1, y_1), (x_2, y_2)) = (d_X(x_1, x_2)^2 + d_Y(y_1, y_2)^2)^{\frac{1}{2}}.$$

Let us denote $[0, \infty)$ by $\overline{\mathbb{R}}_+$. Define the map

$$\rho : (X \times Y) \times (X \times Y) \to \overline{\mathbb{R}}_+ \times \overline{\mathbb{R}}_+$$

by

$$\rho((x_1, y_1), (x_2, y_2)) = (d_X(x_1, x_2), d_Y(y_1, y_2)).$$

Now we define $f : \overline{\mathbb{R}}_+ \times \overline{\mathbb{R}}_+ \to \overline{\mathbb{R}}_+$ by

$$f(x, y) = (x^2 + y^2)^{\frac{1}{2}}.$$

Clearly, $d = f \circ \rho$. We can do the same for an arbitrary product of metric spaces as follows.

Let $\{(X_\alpha, d_\alpha) \mid \alpha \in \Lambda\}$ be an arbitrary family of metric spaces. Let $\delta = (d_\alpha)_{\alpha \in \Lambda}$ be the collection of metrics d_α. Let $Z = \prod_{\alpha \in \Lambda} X_\alpha$, and let $x = (x_\alpha)$ and $y = (y_\alpha)$ be in Z. Define

$$\rho_\delta : Z \times Z \to \overline{\mathbb{R}}_+^\Lambda$$

by

$$\rho_\delta(x, y) = (d_\alpha(x_\alpha, y_\alpha)). \tag{4.2}$$

Let us study the properties of a function $f : A \to \overline{\mathbb{R}}_+$ such that $f \circ \rho_\delta$ is a metric on Z, where $\operatorname{Im} \rho_\delta \subseteq A \subseteq \overline{\mathbb{R}}_+^\Lambda$. Let

$$\sigma : Z \times Z \times Z \to \overline{\mathbb{R}}_+^\Lambda \times \overline{\mathbb{R}}_+^\Lambda \times \overline{\mathbb{R}}_+^\Lambda$$

be the map defined by

$$\sigma(x, y, z) = (\rho_\delta(x, y), \rho_\delta(x, z), \rho_\delta(y, z)).$$

Let us generalize Definitions 2.3.4 and 2.3.6 as follows.

Definition 4.2.1. Let $A \subseteq \prod_{\alpha \in \Lambda} \overline{\mathbb{R}}_+^\Lambda$. A map $f : A \to \overline{\mathbb{R}}_+$ is called
(i) subadditive if for all $x, y \in \prod_{\alpha \in \Lambda} \overline{\mathbb{R}}_+^\Lambda$,

$$f(x + y) \le f(x) + f(y),$$

(ii) amenable on A if $f^{-1}\{0\} = \{(0)\}$.

Theorem 4.2.2 (Borsik and Doboš). *Let $\{(X_\alpha, d_\alpha) \mid \alpha \in \Lambda\}$ be an arbitrary family of metric spaces. Let $\delta = (d_\alpha)_{\alpha \in \Lambda}$ be the collection of metrics d_α. Let ρ_δ be the map as defined in (4.2),* *let A be a subset of $\prod_{\alpha \in \Lambda} \overline{\mathbb{R}}_+^{\Lambda}$ containing the image $\operatorname{Im} \rho$, and let $f : A \to \overline{\mathbb{R}}_+$. Then $f \circ \rho_\delta$ is a metric on $Z = \prod_{\alpha \in \Lambda} X_\alpha$ if and only if*

(i) *f is amenable on $\operatorname{Im} \rho_\delta$,*

(ii) *for all $x, y, z \in \operatorname{Im} \rho_\delta$ such that $(x, y, z) \in \operatorname{Im} \sigma$, we have $f(x) \le f(y) + f(z)$.*

Proof. Suppose that $f \circ \rho_\delta$ is a metric on Z. Let $x \in \operatorname{Im} \rho_\delta$. Then there exist $a, b \in Z$ such that $\rho_\delta(a, b) = x$. Therefore

$$0 = f(x) = f(\rho_\delta(a, b)) \Leftrightarrow a = b \Leftrightarrow x = (0).$$

This proves (i). Now let $x, y, z \in \operatorname{Im} \rho_\delta$ be such that $(x, y, z) \in \operatorname{Im} \sigma$. Then there exist $a, b, c \in Z$ such that

$$\rho_\delta(a, b) = x, \quad \rho_\delta(a, c) = y \quad \text{and} \quad \rho_\delta(b, c) = z.$$

Therefore

$$
\begin{aligned}
f(x) &= f(\rho_\delta(a, b)) \\
&\le f \circ \rho_\delta(a, c) + f \circ \rho_\delta(b, c) \\
&= f(y) + f(z).
\end{aligned}
$$

This proves (ii).

Conversely, suppose that the conditions (i) and (ii) hold. Clearly, $f \circ \rho_\delta(x, y) \ge 0$ and $f \circ \rho_\delta(x, y) = f \circ \rho_\delta(y, x)$. Let $x = (x_\alpha), y = (y_\alpha) \in Z$. Then

$$
\begin{aligned}
0 &= f \circ \rho_\delta(x, y) \\
&= f(\rho_\delta(x, y)) \\
&\Leftrightarrow \rho_\delta(x, y) = (0) \\
&\Leftrightarrow x_\alpha = y_\alpha \text{ for all } \alpha \in \Lambda \\
&\Leftrightarrow x = y.
\end{aligned}
$$

Let $x, y, z \in Z$. Then $\sigma(x, y, z) \in \operatorname{Im} \sigma$. Therefore

$$
\begin{aligned}
f \circ \rho_\delta(x, y) &= f(\rho_\delta(x, y)) \\
&\le f(\rho_\delta(x, z)) + f(\rho_\delta(z, y)) \\
&= f \circ \rho_\delta(x, z) + f \circ \rho_\delta(z, y).
\end{aligned}
$$

This shows that $f \circ \rho_\delta$ is a metric on Z. □

Definition 4.2.3. A map $f : \overline{\mathbb{R}}_+^{\Lambda} \to \overline{\mathbb{R}}_+$ is called a metric-preserving map if $f \circ \rho_\delta$ is a metric for every collection of metrics $\delta = (d_\alpha)_{\alpha \in \Lambda}$.

Proposition 4.2.4. *Let* $f : \overline{\mathbb{R}}_+^\Lambda \to \mathbb{R}_+$ *be a metric-preserving map. Then* f *is amenable.*

Proof. Let $a \in \mathbb{R}_+$. Let d be the usual metric on $\mathbb{R}$. Then there exist $x, y \in \mathbb{R}$ such that $d(x, y) = a$. Define a collection $\delta = (d_\alpha)_{\alpha \in \Lambda}$ of metrics such that $d_\alpha = d$ for every $\alpha \in \Lambda$. Note that $\operatorname{Im} \rho_\delta = \overline{\mathbb{R}}_+^\Lambda$. Then by Theorem 4.2.2 f is amenable. $\qquad\square$

Let $x, y \in \mathbb{R}^\Lambda$. We say that $x \leq y$ if $y - x \in \overline{\mathbb{R}}_+^\Lambda$. Let $x, y, z \in \overline{\mathbb{R}}_+^\Lambda$ be such that $x \leq y + z$, $y \leq z + x$, and $z \leq x + y$. By the abuse of language, we will also call such a triple (x, y, z) a triangle triple.

Proposition 4.2.5. *Let* $f : \overline{\mathbb{R}}_+^\Lambda \to \mathbb{R}_+$ *be a metric-preserving map. If* $(x, y, z) \in \overline{\mathbb{R}}_+^\Lambda \times \overline{\mathbb{R}}_+^\Lambda \times \overline{\mathbb{R}}_+^\Lambda$ *is a triple, then* $f(x) \leq f(y) + f(z)$.

Proof. Let $a, b, c \in \mathbb{R}_+$, and let d be the usual metric on $\mathbb{R}^2$. Then there exist $x, y, z \in \mathbb{R}^2$ such that $d(x, y) = a$, $d(y, z) = b$, and $d(z, x) = c$. Define a collection $\delta = (d_\alpha)_{\alpha \in \Lambda}$ of metrics such that $d_\alpha = d$ for every $\alpha \in \Lambda$. Note that

$$\{(x, y, z) \in \overline{\mathbb{R}}_+^\Lambda \times \overline{\mathbb{R}}_+^\Lambda \times \overline{\mathbb{R}}_+^\Lambda \mid (x, y, z) \text{ is a triangle triple}\} \subseteq \operatorname{Im} \rho_\delta.$$

Then by Theorem 4.2.2, $f(x) \leq f(y) + f(z)$. $\qquad\square$

By Theorem 4.2.2 and Propositions 4.2.4 and 4.2.5 we have the following:

Theorem 4.2.6 (Borsik and Dobos). *Let* $\{(X_\alpha, d_\alpha) \mid \alpha \in \Lambda\}$ *be an arbitrary family of metric spaces. Let* $\delta = (d_\alpha)_{\alpha \in \Lambda}$ *be the collection of metrics* d_α. *Let* ρ_δ *be the map as defined in* (4.2). *Then* $f : \overline{\mathbb{R}}_+^\Lambda \to \mathbb{R}_+$ *is a metric-preserving map if and only if*
(i) f *is amenable,*
(ii) *for all triangle triples* $(x, y, z) \in \overline{\mathbb{R}}_+^\Lambda \times \overline{\mathbb{R}}_+^\Lambda \times \overline{\mathbb{R}}_+^\Lambda$, *we have* $f(x) \leq f(y) + f(z)$.

4.3 Quotient metric space

Definition 4.3.1. Let X be a nonempty set. Then a map $d : X \times X \to \mathbb{R}$ is called a pseudo-metric on X if for all $x, y, z \in X$, the following conditions hold:
(i) $d(x, y) \geq 0$,
(ii) if $x = y$, then $d(x, y) = 0$,
(iii) $d(x, y) = d(y, x)$,
(iv) $d(x, y) \leq d(x, z) + d(z, y)$.

The pair (X, d), where X is a nonempty set and d is a pseudo-metric on X, is called a pseudo-metric space. If (X, d) is a pseudo-metric space, then we can prove that $|d(x, z) - d(y, z)| \leq d(x, y)$ for all $x, y, z \in X$.

Example 4.3.2. Every metric space is a pseudo-metric space. Let X be a set containing at least two elements. Define $d : X \times X \to \mathbb{R}$ by $d(x,y) = 0$ for all $x,y \in X$. Then d is a pseudo-metric but not a metric on X.

Example 4.3.3. Let $R[a, b]$ denote the set of all Riemann-integrable functions $f : [a, b] \to \mathbb{R}$. Define $d : R[a, b] \times R[a, b] \to \mathbb{R}$ by

$$d(f,g) = \int_a^b |f(t) - g(t)| dt.$$

Then d is a pseudo-metric on $R[a, b]$. The value $d(f, g)$ may be 0 without $f = g$, for example, if f and g are two Riemann-integrable functions that differ only at finitely many points of $[a, b]$, then $d(f, g) = 0$.

Example 4.3.4. Let $f : X \to \mathbb{R}$ be a map. Define the map $d : X \times X \to \mathbb{R}$ by $d(x,y) = |f(x) - f(y)|$. Then d is a pseudo-metric on X. We can easily prove that d is a metric if and only if f is injective.

Example 4.3.5. The Hausdorff distance $d_H(A, B)$ defined on the set of all bounded nonempty sets of a metric space is a pseudo-metric.

Proposition 4.3.6. *Let V be a normed linear space, and let W be a subspace of V. Then the map $d_W : V/W \times V/W \to \mathbb{R}$ defined by*

$$d_W(x + W, y + W) = \inf\{\|x - y + w\| \mid w \in W\}$$

is a pseudo-metric on V/W.

Proof. We can clearly observe that $d_W(x + W, y + W) \geq 0$, $d_W(x + W, x + W) = 0$, and $d_W(x + W, y + W) = d_W(y + W, x + W)$. Now we prove the triangle inequality. Observe that

$$d_W(x + W, y + W) = \inf\{\|x - y + w\| \mid w \in W\}$$

$$= \inf\left\{\left\|(x - z) - (y - z) + \frac{w}{2} + \frac{w}{2}\right\| \mid w \in W\right\}$$

$$\leq \inf\left\{\left\|(x - z) + \frac{w}{2}\right\| + \left\|(y - z) + \frac{w}{2}\right\| \mid w \in W\right\}$$

$$\leq \inf\left\{\left\|(x - z) + \frac{w}{2}\right\| \mid w \in W\right\}$$

$$+ \inf\left\{\left\|(y - z) + \frac{w}{2}\right\| \mid w \in W\right\}$$

$$\leq d_W(x + W, z + W) + d_W(z + W, y + W). \qquad \square$$

Remark 4.3.7. We can observe that $d_W(x + W, y + W) = 0$ if and only if $x - y \in Cl\, W$.

Let $T : V \to W$ be a surjective continuous linear transformation, and let $K = \operatorname{Ker} T$. By Proposition 4.3.6 and Remark 4.3.7 we can note that V/K is a normed space. We can also check that the map $\overline{T} : V/K \to W$ defined by

$$\overline{T}(x + K) = T(x)$$

is a bijective linear transformation.

Theorem 4.3.8. *Let V and W be two normed spaces, and let $T : V \to W$ be a surjective continuous linear transformation. Let $K = \operatorname{Ker} T$. Then the map $\overline{T} : V/K \to W$ defined above is a bijective continuous linear transformation, and $\|\overline{T}\| = \|T\|$.*

Proof. Let $y \in x + K$, where $x \in V$. Then

$$\|\overline{T}(x + K)\| = \|T(y)\| \le \|T\| \|y\|.$$

This implies that

$$\|\overline{T}(x + K)\| \le \|T\| \inf\{\|y\| \mid y \in x + K\}$$
$$= \|T\| \|x + K\|.$$

This shows that $\overline{T}$ is continuous. Now note that

$$\|\overline{T}\| = \sup\{\|\overline{T}(x + K)\| \mid \|x + K\| \le 1\} \le \|T\|.$$

Let $x \in V$ be such that $\|x\| \le 1$. Then

$$\|x + K\| = \inf\{\|x + z\| \mid z \in K\} \le \|x\| \le 1.$$

Therefore

$$\|T(x)\| = \|\overline{T}(x + K)\|$$
$$\le \sup\{\|\overline{T}(w + K)\| \mid \|w + K\| \le 1\}$$
$$= \|\overline{T}\|.$$

This shows that

$$\|T\| = \sup\{\|T(x)\| \mid \|x\| \le 1\} \le \|\overline{T}\|.$$

Thus $\|\overline{T}\| = \|T\|$. □

Let (X, d) be a pseudo-metric space. Define a relation $\sim$ on X by $x \sim y$ if $d(x, y) = 0$. Then we can check that $\sim$ is an equivalence relation on X.

Proposition 4.3.9. *Let (X, d) be a pseudo-metric space, and let $\sim$ be the equivalence relation defined above. Let $\overline{X} = X / \sim$. Then the map $\tilde{d} : \overline{X} \times \overline{X} \to \mathbb{R}$ defined by $\tilde{d}(\overline{x}, \overline{y}) = d(x, y)$ is a metric on $\overline{X}$.*

Proof. We first show that $\tilde{d}$ is well defined. Let $x_1, x_2 \in \overline{x}$ and $y_1, y_2 \in \overline{y}$. Then

$$d(x_1, y_1) \leq d(x_1, x_2) + d(x_2, y_2) + d(y_2, y_1) = d(x_2, y_2),$$

and

$$d(x_2, y_2) \leq d(x_2, x_1) + d(x_1, y_1) + d(y_1, y_2) = d(x_1, y_1).$$

This shows that $d(x_1, y_1) = d(x_2, y_2)$. Suppose that $\tilde{d}(\overline{x}, \overline{y}) = 0$. Therefore $d(x, y) = 0$. This implies that $x \sim y$, which shows that $\overline{x} = \overline{y}$. We can easily show that the other conditions of a metric are satisfied. □

We will write X/d instead of $(X/ \sim, \tilde{d})$ and use the same symbol d in place of $\tilde{d}$ to denote the metric on X/d.

Let (X, d) be a metric space, and let R be an equivalence relation on the set X. Define the map $d_R : X \times X \to \mathbb{R}$ by

$$d_R(\overline{x}, \overline{y}) = \inf\{d(p_0, q_0) + d(p_1, q_1) + \cdots + d(p_n, q_n)\},$$

where the infimum is taken over all the finite sequences $(p_0, p_1, \ldots, p_n)$ and $(q_0, q_1, \ldots, q_n)$ with $\overline{p_0} = \overline{x}, \overline{q_n} = \overline{y}$, and $\overline{q_i} = \overline{p_{i+1}}$ for all $i = 1, 2, \ldots, n-1$. We can check that d_R satisfies the first three conditions of a pseudo-metric space. We now prove that it also satisfies the triangle inequality. Let $\overline{x}, \overline{y}, \overline{z} \in X/R$. Let $p_0, p_1, \ldots, p_n, q_0, q_1, \ldots, q_n, p'_0, p'_1, \ldots, p'_m$, and $q'_0, q'_1, \ldots, q'_m$ be such that $\overline{p_0} = \overline{x}, \overline{q_n} = \overline{y} = \overline{p'_0}, \overline{q'_m} = \overline{z}, \overline{q_i} = \overline{p_{i+1}}$, and $\overline{q'_j} = \overline{p'_{j+1}}$ for all i and j. Then

$$\sum_{i=0}^{n} d(p_i, q_i) + \sum_{j=0}^{m} d(p'_j, q'_j) = \sum_{i=0}^{n-1} d(p_i, q_i) + d(p_n, q_n) + d(p'_0, q'_0) + \sum_{j=1}^{m} d(p'_j, q'_j)$$

$$\geq \sum_{i=0}^{n-1} d(p_i, q_i) + d(p_n, q'_0) + \sum_{j=1}^{m} d(p'_j, q'_j).$$

Considering the infimum over all such p_i and q_j, we get

$$d_R(\overline{x}, \overline{y}) + d_R(\overline{y}, \overline{z}) \geq d_R(\overline{x}, \overline{z}).$$

Thus d_R is a pseudo-metric on X/R. By Proposition 4.3.9 we can define a metric on the quotient of X/R, which we call the quotient metric. We denote it by the same notation $(X/R, d_R)$. The space $(X/R, d_R)$ is called the quotient metric space.

Definition 4.3.10. Let (X, d_X) and (Y, d_Y) be pseudo-metric spaces. Then a map $f : X \to Y$ is called nonexpansive if for all $x, y \in X$,

$$d_Y(f(x), f(y)) \le d_X(x, y).$$

We can check that the pseudo-metric d_R defined above satisfies the following universal property:

If $f : X \to Y$ is a nonexpansive map such that $f(x) = f(y)$ whenever $\overline{x} = \overline{y}$, then there is a nonexpansive map $\overline{f} : X/R \to Y$ such that the following diagram commutes:

$$X \xrightarrow{\ \nu\ } X/R$$

$$\begin{array}{c} f \searrow \quad \downarrow \overline{f} \\ Y \end{array}$$

where ν is the quotient map.

Let V be a normed space, and let W be a subspace of V. Then the relation R on V defined by $(x, y) \in R$ if $x - y \in W$ is an equivalence relation V. Observe that for each $x \in V$, the equivalence class $\overline{x} = x + W$. We now observe that $d_R = d_W$.

Let $p_0, p_1, \ldots, p_n$ be such that $\overline{p_0} = \overline{x}$, $\overline{q_n} = \overline{y}$, and $\overline{q_i} = \overline{p_{i+1}}$ for all i. Then

$$\sum_{i=0}^{n} d(p_i, q_i) = \sum_{i=0}^{n} \|(p_i - q_i)\|$$

$$= \|p_0 - q_0\| + \|p_1 - q_1\| + \cdots + \|p_n - q_n\|$$

$$= \|p_0 - q_0\| + \|q_0 + w_1 - q_1\| + \cdots + \|q_{n-1} + w_n - q_n\|$$

$$\ge \|p_0 - q_n + (w_1 + \cdots + w_n)\|.$$

This shows that

$$d_R(\overline{x}, \overline{y}) \ge d_W(\overline{x}, \overline{y}).$$

Note that

$$\{\|x - y + w\| \mid w \in W\} = \{\|p - q\| \mid \overline{p} = \overline{q}\}$$

$$\subseteq \left\{ \sum_{i=0}^{n} \|(p_i - q_i)\| \ \middle|\ \overline{p_0} = \overline{x}, \overline{q_n} = \overline{y}, \overline{q_i} = \overline{p_{i+1}}, \forall i \right\}.$$

This shows that

$$d_R(\overline{x}, \overline{y}) \le d_W(\overline{x}, \overline{y}).$$

Hence $d_R(\overline{x}, \overline{y}) = d_W(\overline{x}, \overline{y})$.

Exercises

4.1. First of all, complete whatever is left for you as exercises.

4.2. Let $\{(X_i, d_i) \mid i \in \mathbb{N}\}$ be a bounded countable family of metric spaces. Consider the metric d on $\prod_{i=1}^{\infty} X_i$ as defined in equation (4.1). Let Y be a metric space. Then show that a map $f : Y \to \prod_{i=1}^{\infty} X_i$ is continuous if and only if $\pi_j \circ f : Y \to X_j$ is continuous for all j, where π_j is the jth projection.

4.3. Let $\{(X_\alpha, d_\alpha) \mid \alpha \in \Lambda\}$ be an arbitrary family of metric spaces. Consider the uniform metric on $\prod_{\alpha \in \Lambda} X_\alpha$. Check whether the statement of Exercise 4.2. holds in this case or not.

4.4. Let $(\prod_{i=1}^{\infty} X_i, d_i)$ be the metric space as in Exercise 4.2. Show that each projection is an open map.

4.5. Let $(\prod_{\alpha \in \Lambda} X_\alpha, d)$ be the metric space as in Exercise 4.3. Check whether the statement of Exercise 4.4. holds in this case or not.

4.6. Consider a group $(G, \circ)$ that is also a metric space. Suppose that $x \mapsto x^{-1}$ is a continuous map on G. Further, consider a metric on $G \times G$ such that the binary operation $\circ$ is continuous. Let U be an open set containing the identity e of G. Show that U contains an open set V containing e such that $V^2 \subseteq U$ and $V^{-1} = V$, where $V^2 = \{xy \mid x, y \in V\}$ and $V^{-1} = \{x^{-1} \mid x \in V\}$.

4.7. Consider a group $(G, \circ)$ that is also a metric space. Suppose that all the hypotheses of Exercise 4.6. hold. Show that the closure of each subgroup is also a subgroup. Also, show that if H is an open subgroup, then it is also a closed subgroup of G.

4.8. Show that $d'((x_1, y_1), (x_2, y_2)) = |(x_1 - x_2) + (y_1 - y_2)|$ is a pseudo-metric on $\mathbb{R}^2$. Also, show that $\mathbb{R}^2/d$ is isometric to the real line.

4.9. Consider $[0, 1]$ with the usual metric d. Let R be the equivalence relation on $[0, 1]$ defined by $(0, 1) \in R$ with the other elements related to themselves only. Explore the quotient metric space $([0, 1]/R, d_R)$.

4.10. Let $f : \overline{\mathbb{R}_+^\Lambda} \to \mathbb{R}_+$ be a metric-preserving map. Show that f is continuous if and only if f is continuous at $0 \in \overline{\mathbb{R}_+^\Lambda}$.

5 Sequences in metric spaces

In this chapter, we study the convergence of sequences in a metric space. We also study the complete metric space.

5.1 Convergent sequences

Recall that a sequence in a metric space X is a map $f : \mathbb{N} \to X$. If $f(n) = x_n$, then x_n is called the nth term of the sequence. We denote a sequence by (x_n) or

$$x_1, x_2, \ldots, x_n, \ldots .$$

Example 5.1.1. Let $c \in X$. Then the map $f : \mathbb{N} \to X$ defined by $f(n) = c$ for all $n \in \mathbb{N}$ is called a constant sequence in X.

Example 5.1.2. Let $m \in \mathbb{N}$ and $c \in X$. Let (x_n) be a sequence in X such that $x_n = c$ for all $n > m$. We say that this sequence is constant on the tail.

Definition 5.1.3. A sequence (x_n) in a metric space X is said to converge to a point x in X if each open ball $B(x, \epsilon)$ around x contains all but finitely many terms of the sequence (x_n) or, equivalently, if for each $\epsilon > 0$, there exists $k \in \mathbb{N}$ such that

$$n \geq k \implies d(x_n, x) < \epsilon.$$

If a sequence (x_n) in X converges to $x \in X$, then we say that (x_n) is a convergent sequence.

If a sequence (x_n) converges to $x \in X$, then we also say that x is the limit of (x_n) or x_n tends to x. We denote this by

$$\lim_{n \to \infty} x_n = x \quad \text{or} \quad x_n \to x.$$

A sequence (x_n) in X that is not convergent is called a divergent sequence. Observe from the definition that if a sequence (x_n) in X converges to $x \in X$, then the sequence $(d(x_n, x))$ converges to 0 in the real line.

Example 5.1.4. A constant sequence (c) in a metric space X is convergent, since for any $\epsilon > 0$, the open ball $B(c, \epsilon)$ contains all the terms of the sequence.

Example 5.1.5. Consider the sequence $(\frac{1}{n})$ in the real line $\mathbb{R}$. Let $\epsilon > 0$. By the Archimedean property there exists $k \in \mathbb{N}$ such that $k > \frac{1}{\epsilon}$. This implies that

$$n \geq k \implies \frac{1}{n} \leq \frac{1}{k} < \epsilon.$$

In other words,

https://doi.org/10.1515/9783111636085-005

$$n \geq k \Rightarrow \left| \frac{1}{n} - 0 \right| < \epsilon.$$

Therefore $\frac{1}{n} \to 0$.

Example 5.1.6. The sequence $\left(\frac{1}{n}\right)$ is not convergent in $(0, 1]$ with the metric induced by that of the real line. On the contrary, suppose that it converges to $x \in (0, 1]$. Then for $\epsilon = \frac{x}{2}$, there is $k \in \mathbb{N}$ such that

$$n \geq k \Rightarrow \left| \frac{1}{n} - x \right| < \frac{x}{2}.$$

This implies that

$$n \geq k \Rightarrow \frac{x}{2} < \frac{1}{n}.$$

This is a contradiction.

Example 5.1.7. Let (x_n) be a convergent sequence in the discrete metric space X such that $x_n \to x \in X$. For $\epsilon = \frac{1}{2}$, there is $k \in \mathbb{N}$ such that

$$n \geq k \Rightarrow d(x_n, x) < \frac{1}{2}.$$

This shows that (x_n) is constant on the tail.

Example 5.1.8. The sequence (n) in the real line $\mathbb{R}$ is divergent. On the contrary, suppose that (n) converges to some $x \in \mathbb{R}$. Then by the Archimedean property there exists $k \in \mathbb{N}$ such that $k > x$. Then for $n \geq k$, $n > x$. Let $0 < \epsilon < k - x$. Then the open interval $(x - \epsilon, x + \epsilon)$ contains no point of the sequence (n), a contradiction.

Example 5.1.9. Consider the sequence $(-1)^n$ in the real line $\mathbb{R}$. Note that the open interval $(1 - \frac{1}{2}, 1 + \frac{1}{2})$ leaves infinitely many terms of the sequence (namely, -1) outside this interval. Therefore 1 can not be the limit of this sequence. Similarly, -1 cannot be the limit of this sequence. Let $x \in \mathbb{R} \setminus \{1, -1\}$. Let $\epsilon = \min\{|1 - x|, |-1 - x|\}$. Then the interval $(x - \epsilon, x + \epsilon)$ contains no terms of this sequence. Hence $(-1)^n$ is divergent in the real line $\mathbb{R}$.

Example 5.1.10. Let (x_n^i) be a sequence in a metric space (X_i, d_i) converging to x_i in X_i, where $1 \leq i \leq m$. Let $Z = \prod_{i=1}^{m} X_i$, and let d be the metric on Z defined by

$$d(a, b) = \left(\sum_{i=1}^{m} d_i(a_i, b_i)^2 \right)^{\frac{1}{2}}$$

for $a = (a_1, \ldots, a_m)$ and $b = (b_1, \ldots, b_m)$ in Z. Let $\epsilon > 0$. Then there are $k_i \in \mathbb{N}$ $(i \leq m)$ such that

$$n \geq k_i \Rightarrow d_i(x_n^i, x_i) < \frac{\epsilon}{\sqrt{m}}.$$

Let $k = \max\{k_i, \ldots, k_m\}$. Then

$$n \geq k \Rightarrow d_i(x_n^i, x_i) < \frac{\epsilon}{\sqrt{m}}.$$

Now for $n \geq k$, we have

$$d((x_n^1, \ldots, x_n^m), (x_1, \ldots, x_m)) = \left(\sum_{i=1}^{m} d_i(x_n^i, x_i)^2 \right)^{\frac{1}{2}}$$

$$< \left(\sum_{i=1}^{m} \frac{\epsilon^2}{m} \right)^{\frac{1}{2}}$$

$$= \epsilon.$$

This shows that the sequence $(x_n^1, \ldots, x_n^m)$ converges to $(x_1, \ldots, x_m)$ in (Z, d).

Let ρ be a metric on Z that is Lipschitz equivalent to d. Then the sequence $(x_n^1, \ldots, x_n^m)$ converges to $(x_1, \ldots, x_m)$ in (Z, ρ).

Proposition 5.1.11. *The limit of a convergent sequence in a metric space X is unique.*

Proof. Let a sequence (x_n) in X converge to x and y in X. If $x \neq y$, then $\epsilon = |x-y| > 0$. Since $x_n \to x$, the open ball $B(x, \frac{\epsilon}{2})$ contains all but finitely many terms of the sequence (x_n). In this case, the open ball $B(y, \frac{\epsilon}{2})$ will contain only finitely many terms of the sequence (x_n). This is a contradiction. □

Definition 5.1.12. Let $f : \mathbb{N} \to X$ be a sequence in a metric space X, and let $i : \mathbb{N} \to \mathbb{N}$ be a strictly increasing sequence map. Then $f \circ i : \mathbb{N} \to X$ is called a subsequence of the sequence f. We denote the subsequence of a sequence (x_n) by (x_{i_n}), where $i_n = i(n)$.

Example 5.1.13. Consider the sequence (x_n) in $\mathbb{R}$ with $x_n = n^2$, that is, the following sequence:

$$1, 4, 9, 16, 25, \ldots.$$

Let $i : \mathbb{N} \to \mathbb{N}$ be the map defined by $i(n) = 2n - 1$. Then the subsequence (x_{i_n}) of the sequence (x_n) is the following:

$$1, 9, 25, \ldots.$$

Proposition 5.1.14. *A subsequence of a convergent sequence in a metric space is convergent.*

Proof. Let (x_n) be a sequence in a metric space X such that $x_n \to x$ in X. Let (x_{i_n}) be a subsequence of (x_n). We will show that $x_{i_n} \to x$. Let $\epsilon > 0$. Since $x_n \to x$, there is $k \in \mathbb{N}$ such that

$$n \geq k \Rightarrow d(x_n, x).$$

We can inductively observe that $i_n \geq n$ for all $n \in \mathbb{N}$. Now

$$i_n \geq n \Rightarrow i_n \geq k.$$

Therefore, for $i_n \geq k$, we have $d(x_{i_n}, x) < \epsilon$. This shows that $x_{i_n} \to x$. $\qquad\square$

A divergent sequence may have a convergent subsequence. For example, the sequence $(-1)^n$ is divergent, but it has a constant subsequence (1).

Proposition 5.1.15. *A set A in a metric space X is closed if and only if every sequence in A that converges in X has its limit in A.*

Proof. Let A be closed in X, and let (x_n) be a sequence in A such that $x_n \to x \in X$. We claim that $x \in A$. On the contrary, suppose that $x \in X \setminus A$. Since $X \setminus A$ is open, there is an open ball $B(x, \epsilon)$ contained in $X \setminus A$. Since the sequence (x_n) lies in A, $B(x, \epsilon)$ contains no term of the sequence. This is a contradiction.

Conversely, suppose that every sequence in A that converges in X has its limit in A. If the derived set $D(A)$ is the empty set, then A is closed. Suppose $D(A) \neq \emptyset$ and $x \in D(A)$. Then for each $n \in \mathbb{N}$, the open ball $B(x, \frac{1}{n})$ contains a point x_n of A other than x. In turn, we get a sequence (x_n) in A. We claim that $x_n \to x$.

Let $\epsilon > 0$. By the Archimedean property of $\mathbb{R}$, there is $k \in \mathbb{N}$ such that $\frac{1}{k} < \epsilon$. Then

$$x_k \in B\left(x, \frac{1}{k}\right) \subseteq B(x, \epsilon).$$

Also, for $n \geq k$, we have

$$x_n \in B\left(x, \frac{1}{n}\right) \subseteq B\left(x, \frac{1}{k}\right) \subseteq B(x, \epsilon).$$

Hence $x_n \to x$. By the assumption, $x \in A$. Since $x \in D(A)$ is arbitrary, $D(A) \subseteq A$. This shows that A is closed. $\qquad\square$

Example 5.1.16. We can use Proposition 5.1.15 to show that $(0, 1]$ is not closed in $\mathbb{R}$. The sequence $(\frac{1}{n})$ in $(0, 1]$ converges to 0 in $\mathbb{R}$ but not in $(0, 1]$.

Proposition 5.1.17. *Let X be a metric space, and let $A \subseteq X$. Then $x \in X$ is a limit point of A if and only if there is a sequence (x_n) of distinct terms in A such that $x_n \to x$.*

Proof. Let $x \in X$ be a limit point of A. Then there is a point $x_1 \in B(x,1) \cap A$ other than x. Suppose that we have inductively obtained n distinct terms $x_1, \ldots, x_n$ other than x. Since the derived set of the set $K = \{x_1, \ldots, x_n\}$ is empty, K is closed in X. Therefore

$$B\left(x, \frac{1}{n+1}\right) \cap (X \setminus K)$$

is open in X and contains x. Since x is a limit point of A, there is

$$x_{n+1} \in B\left(x, \frac{1}{n+1}\right) \cap (X \setminus K) \cap A$$

other than x. By an argument similar to that in the proof of Proposition 5.1.15 we get $x_n \to x$.

Conversely, suppose that (x_n) is a sequence of distinct terms in A such that $x_n \to x$. Let $\epsilon > 0$. Then there is $k \in \mathbb{N}$ such that

$$n \geq k \Rightarrow x_n \in B(x, \epsilon).$$

This shows that $B(x, \epsilon)$ contains a point of A other than x. Hence x is a limit point of A. □

Proposition 5.1.18. *Let X and Y be metric spaces. Then a map $f : X \to Y$ is continuous at $x \in X$ if and only if for every sequence (x_n) in X such that $x_n \to x$, we have $f(x_n) \to f(x)$.*

Proof. Let $f : X \to Y$ be a continuous map at $x \in X$, and let (x_n) be a sequence in X such that $x_n \to x$. Choose $\epsilon > 0$ and an open ball $B(f(x), \epsilon)$ around $f(x)$. Since f is continuous at x, there is $\delta > 0$ such that

$$f(B(x, \delta)) \subseteq B(f(x), \epsilon). \tag{5.1}$$

Since $x_n \to x$, $B(x, \delta)$ contains all but finitely many terms of the sequence (x_n). Let

$$x_{r_1}, x_{r_1}, \ldots, x_{r_m}$$

be the terms of the sequence (x_n) that are outside the open ball $B(x, \delta)$. From equation (5.1) we see that there are at most m terms of the sequence $(f(x_n))$ that are outside the open ball $B(f(x), \epsilon)$. This shows that $f(x_n) \to f(x)$.

Conversely, suppose that $f(x_n) \to f(x)$ as $x_n \to x$. On the contrary, suppose that f is not continuous at x. Then there is $\epsilon > 0$ such that for all $\delta > 0$, we have

$$d(y,x) < \delta \quad \text{but} \quad d(f(y), f(x)) \geq \epsilon.$$

In other words,

$$y \in B(x, \delta) \quad \text{but} \quad f(y) \notin B(f(x), \epsilon).$$

For each $n \in \mathbb{N}$, choose $x_n \in B(x, \frac{1}{n})$. By an argument similar to that in the proof of Proposition 5.1.15 we get $x_n \to x$. By the assumption, $f(x_n) \to f(x)$. This is a contradiction as $f(x_n) \notin B(f(x), \epsilon)$ for $n \in \mathbb{N}$ such that $\frac{1}{n} < \delta$. Hence f is continuous at x. $\square$

We may use Proposition 5.1.18 in showing that a certain function is not continuous at a certain point.

Example 5.1.19. Consider the map $f : \mathbb{R} \to \mathbb{R}$ defined by

$$f(x) = \begin{cases} 0 & \text{if } x \in \mathbb{Q}, \\ 1 & \text{if } x \in \mathbb{R} \setminus \mathbb{Q}. \end{cases}$$

Let $x \in \mathbb{R}$ be a rational number. Choose a sequence (x_n) of irrational numbers such that $x_n \to x$. Then $f(x_n) = 1$ for all $n \in \mathbb{N}$, but $f(x_n) \not\to f(x) = 0$. Hence f is not continuous at the rational numbers. Similarly, f is not continuous at irrational numbers.

Example 5.1.20. Consider the map $f : \mathbb{R}^2 \to \mathbb{R}$ defined by

$$f(x,y) = \begin{cases} \frac{xy}{x^2+y^2} & \text{if } (x,y) \neq (0,0), \\ 0 & \text{if } (x,y) = (0,0). \end{cases}$$

Consider the sequence $(\frac{1}{n}, \frac{1}{n})$ in $\mathbb{R}^2$. Note that $(\frac{1}{n}, \frac{1}{n}) \to (0,0)$ and

$$f\left(\frac{1}{n}, \frac{1}{n}\right) = \frac{n}{n+1} \to 1.$$

Since $f(0,0) = 0$, f is not continuous at $(0,0)$.

Let (X, d) be a metric space. Let $x, y, a, b \in X$. By the triangle inequality we have

$$d(x,y) \leq d(x,a) + d(a,b) + d(b,y).$$

This implies that

$$d(x,y) - d(a,b) \leq d(x,a) + d(b,y).$$

By a similar argument we have

$$d(a,b) - d(x,y) \leq d(x,a) + d(b,y).$$

Therefore we have

$$|d(x,y) - d(a,b)| \leq d(x,a) + d(b,y). \tag{5.2}$$

Suppose $X \times X$ is provided with a metric such that $(x_n, y_n) \to (x,y)$ implies that $x_n \to x$ and $y_n \to y$, for example,

$$d'((x_1, y_1), (x_2, y_2)) = d(x_1, x_2) + d(y_1, y_2).$$

By equation (5.2) we have

$$|d(x_n, y_n) - d(x, y)| \leq d(x_n, x) + d(y_n, y).$$

If $(x_n, y_n) \to (x, y)$, then $d(x_n, y_n) \to d(x, y)$. This shows that under such an assumption, the metric d is continuous map from $X \times X$ to $\mathbb{R}$.

Remark 5.1.21. If $X \times X$ is provided with a metric such that $(x_n, y_n) \to (x, y)$ does not imply that $x_n \to x$ and $y_n \to y$, then the metric $d : X \times X \to \mathbb{R}$ need not by continuous. For example, consider a bijective map $f : X \to X$ which is not continuous such that $x_n \to x$ but $f(x_n) \to u = f(v) \neq f(x)$. Define the metric

$$d' : X \times X \to \mathbb{R}$$

by

$$d'((x_1, y_1), (x_2, y_2)) = d(f(x_1), f(x_2)) + d(f(y_1), f(y_2)).$$

Now

$$d'((x_n, 0), (v, 0)) = d(f(x_n), f(v)).$$

Then $(x_n, 0) \to (v, 0)$, but $x_n \nrightarrow v$.

Remark 5.1.22. In the forthcoming, we will discuss the product topology. The above remark implies that the product topology is different from the topology induced by such a metric.

Let us consider the following proposition, whose proof is left as exercise.

Proposition 5.1.23. *Let (x_n) and (y_n) be sequences in $\mathbb{R}^m$ such that $x_n \to x$ and $y_n \to y$. Then*
(i) $x_n + y_n \to x + y$,
(ii) $x_n y_n \to xy$,
(iii) $a x_n \to ax$ *for $a \in \mathbb{R}$.*

As an application of Propositions 5.1.18 and 5.1.23, we have the following:

Corollary 5.1.24. *The binary operation of usual addition, pointwise multiplication from $\mathbb{R}^m \times \mathbb{R}^m$ to $\mathbb{R}^m$, and the scalar multiplication from $\mathbb{R} \times \mathbb{R}^m$ to $\mathbb{R}^m$ are continuous maps.*

From the inequality

$$\big|\|x\| - \|y\|\big| \leq \|x - y\|$$

we see that if $x_n \to x$ in a normed space V, then $\|x_n\| \to \|x\|$ in $\mathbb{R}$. This shows that a norm is a continuous map from V to $\mathbb{R}$.

Proposition 5.1.25. *Let d_1 and d_2 be Lipschitz equivalent metrics on a nonempty set X. Then a sequence (x_n) is convergent in (X, d_1) if and only if (x_n) is convergent in (X, d_2).*

Proof. Since d_1 and d_2 are Lipschitz equivalent, there are positive constants m and M such that

$$md_1(x, y) \le d_2(x, y) \le Md_1(x, y)$$

for all $x, y \in X$. Let x_n be a sequence in (X, d_1) such that x_n converges to x in (X, d_1). Let $\epsilon > 0$. Then there is $k \in \mathbb{N}$ such that

$$n \ge k \Rightarrow d_1(x_n, x) < \frac{\epsilon}{M}.$$

Therefore $d_2(x_n, x) < Md_1(x, y) < \epsilon$. This shows that $x_n \to x$ in (X, d_2). We can similarly show the converse. $\qquad\square$

Proposition 5.1.26. *Let d_1 and d_2 be metrics on a nonempty set X such that a sequence (x_n) is converges to x in (X, d_1) if and only if (x_n) is converges to x in (X, d_2). Then d_1 and d_2 are topologically equivalent.*

Proof. We will show that the closed sets in (X, d_1) and (X, d_2) are precisely same. Let F be a closed set in (X, d_1). Let $x \in X$ be a limit point of F in (X, d_2). Then there is a sequence (x_n) of distinct terms such that $x_n \to x$ in (X, d_2). By the assumption, $x_n \to x$ in (X, d_1). This shows that $x \in X$ is a limit point of F in (X, d_1). Since F is closed in (X, d_1), $x \in F$. Therefore F is closed in (X, d_2). Similarly, a closed set in (X, d_2) is closed in (X, d_1). $\qquad\square$

Example 5.1.27. We have seen that the metric spaces (X, d) and $(X, \overline{d})$ are topologically equivalent, where

$$\overline{d}(x, y) = \frac{d(x, y)}{1 + d(x, y)}.$$

Let us use Proposition 5.1.26 to show this fact. Let (x_n) be a sequence in X such that $x_n \to x$ in (X, d). This implies that $d(x_n, x) \to 0$. Then

$$\overline{d}(x_n, x) = \frac{d(x_n, x)}{1 + d(x_n, x)}$$

$$\le d(x_n, x) \to 0.$$

Therefore $x_n \to x$ in $(X, \overline{d})$.

Conversely, suppose that $x_n \to x$ in $(X, \overline{d})$. This implies that $\overline{d}(x_n, x) \to 0$. We claim that the set $\{d(x_n, x) \mid n \in \mathbb{N}\}$ is bounded. On the contrary, suppose that $\{d(x_n, x) \mid n \in \mathbb{N}\}$

is unbounded. Then for each real number $M > 0$, there is $k \in \mathbb{N}$ such that $d(x_{n_k}, x) > M$. This will imply that

$$\overline{d}(x_n, x) = \frac{d(x_n, x)}{1 + d(x_n, x)} \rightarrow 1.$$

This is a contradiction. Suppose $d(x_n, x)$ is bounded by some real number $L > 0$. Then

$$\overline{d}(x_n, x) = \frac{d(x_n, x)}{1 + d(x_n, x)} \geq \frac{d(x_n, x)}{1 + L}.$$

This implies that

$$d(x_n, x) \leq (1 + L)\overline{d}(x_n, x) \rightarrow 0.$$

This shows that $x_n \rightarrow x$ in $(X, \overline{d})$.

5.2 Complete metric spaces

In the last section, we defined a convergent sequence. To show that a sequence is convergent, we need to know the point to which it is convergent. Sometimes, it may not be easy to know to which point a sequence is going to converge. We saw that if a sequence (x_n) converges to a point x, then the terms of the sequence are getting closure to x after some kth term. We will see that if a sequence is convergent, then the terms of the sequence will get closure after some kth term. Augustin-Louis Cauchy defined the convergent sequence for the real numbers in this way, which we now call a Cauchy sequence. We will see that a convergent sequence is not exactly the same as a Cauchy sequence.

Definition 5.2.1. A sequence (x_n) is a metric space (X, d) is called a Cauchy sequence if for each $\epsilon > 0$, there exists $k \in \mathbb{N}$ such that

$$m, n \geq k \Rightarrow d(x_m, x_n) < \epsilon.$$

Example 5.2.2. Let (x_n) be a Cauchy sequence in the discrete metric space (X, d). Then for $\epsilon = 1$, there is $k \in \mathbb{N}$ such that

$$m, n \geq k \Rightarrow d(x_m, x_n) < 1.$$

This shows that (x_n) is constant on tail.

Proposition 5.2.3. *A convergent sequence in a metric space is a Cauchy sequence.*

Proof. Let (x_n) be a convergent sequence in a metric space (X, d) such that $x_n \rightarrow x$ in X. Let $\epsilon > 0$. Then there is $k \in \mathbb{N}$ such that

$$n \geq k \Rightarrow d(x_n, x) < \frac{\epsilon}{2}.$$

Let $m, n \geq k$. Then $d(x_m, x) < \frac{\epsilon}{2}$ and $d(x_n, x) < \frac{\epsilon}{2}$. By the triangle inequality we have

$$d(x_m, x_n) \leq d(x_m, x) + d(x_n, x) < \frac{\epsilon}{2} + \frac{\epsilon}{2} = \epsilon.$$

Hence (x_n) is a Cauchy sequence, □

Example 5.2.4. A Cauchy sequence need not be convergent. Consider the subspace $(0, 1]$ of $\mathbb{R}$ with the usual metric. Let $\epsilon > 0$. By the Archimedean property there is $k \in \mathbb{N}$ such that $\frac{1}{k} < n$. For all $m, n \geq k$ with $n \geq m$, we have

$$\left| \frac{1}{m} - \frac{1}{n} \right| = \frac{1}{m} - \frac{1}{n} < \frac{1}{m} < \epsilon.$$

This shows that the sequence $(\frac{1}{n})$ is Cauchy. We have seen in Example 5.1.6 that $(\frac{1}{n})$ is not convergent in $(0, 1]$.

Definition 5.2.5. A sequence (x_n) in a metric space is called a bounded sequence if the set $\{x_n \mid n \in \mathbb{N}\}$ is a bounded set in X.

Example 5.2.6. The sequences $((-1)^n)$ and $(\frac{1}{n})$ are bounded sequences in $\mathbb{R}$, whereas the sequence (n^2) is not a bounded sequence in $\mathbb{R}$.

Proposition 5.2.7. *A Cauchy sequence is a bounded sequence.*

Proof. Let (x_n) be a Cauchy sequence in a metric space X. For $\epsilon = 1$, there is $k \in \mathbb{N}$ such that

$$m, n \geq k \Rightarrow d(x_m, x_n) < 1.$$

Let

$$M = \max\{d(x_1, x_k), d(x_2, x_k), \ldots, d(x_{k-1}, x_k), 1\}.$$

Note that $d(x_n, x_k) < M + 1$ for all $n \in \mathbb{N}$. Hence (x_n) is a bounded sequence. □

Corollary 5.2.8. *A convergent sequence is a bounded sequence.*

Example 5.2.9. A bounded sequence need not be Cauchy. For example, $((-1)^n)$ is bounded in $\mathbb{R}$ but not Cauchy.

The following result characterizes when a bounded sequence is a Cauchy sequence.

Proposition 5.2.10. *Let (x_n) be a bounded sequence in a metric space X. For each $n \in \mathbb{N}$, let $E_n = \{x_l \mid l \geq n\}$. Then (x_n) is Cauchy if and only if the sequence $(\mathrm{diam}(E_n))$ in $\mathbb{R}$ converges to 0.*

Proof. Suppose (x_n) is a Cauchy sequence. Let $\epsilon > 0$. Then there is $k \in \mathbb{N}$ such that

$$m, n \geq k \Rightarrow d(x_m, x_n) < \frac{\epsilon}{2}.$$

Note that $x_m, x_n \in E_k$. Then $\operatorname{diam}(E_k) \leq \frac{\epsilon}{2}$. For $n \geq k$, $E_n \subseteq E_k$. Then

$$n \geq k \Rightarrow |\operatorname{diam}(E_n) - 0| \leq \frac{\epsilon}{2} < \epsilon.$$

This implies that $\operatorname{diam}(E_n) \to 0$.

Conversely, suppose that $\operatorname{diam}(E_n) \to 0$. Let $\epsilon > 0$. Then there is $k \in \mathbb{N}$ such that

$$n \geq k \Rightarrow |\operatorname{diam}(E_n) - 0| < \epsilon.$$

This implies that $d(x_m, x_n) < \epsilon$ for $m, n \geq k$, which shows that (x_n) is a Cauchy sequence. $\qquad\square$

Definition 5.2.11. A metric space (X, d) is called a complete metric space is every Cauchy sequence in X is convergent. If (X, d) is a complete metric space, then d is called a complete metric on X.

We say that a normed space is complete if the metric induced by the norm is complete. Also, we say that an inner product space is complete if the norm induced by the inner product is complete. A complete normed space is called a Banach space, and a complete inner product space is called a Hilbert space.

Example 5.2.12. The discrete metric space is complete.

Example 5.2.13. The space $(0, 1)$ as a subspace of the real line $\mathbb{R}$ is not complete.

Remark 5.2.14. We can define a metric on $(0, 1)$ that is topologically equivalent to the usual metric on $(0, 1)$ and is complete with respect to the new metric. For example, define the metric ρ on $(0, 1)$ by

$$\rho(x, y) = |x - y| + \left| \frac{1}{m(x)} - \frac{1}{m(y)} \right|,$$

where $m(x) = \min\{x, 1 - x\}$. Then ρ is complete on $(0, 1)$ and topologically equivalent to the usual metric on $(0, 1)$.

Example 5.2.15. Consider the real line $\mathbb{R}$. Define the map $f : \mathbb{R} \to \mathbb{R}$ by

$$f(x) = \frac{x}{1 + |x|}.$$

Define the metric d on $\mathbb{R}$ by $d(x, y) = |f(x) - f(y)|$. Note that f is a homeomorphism between $\mathbb{R}$ and $(-1, 1)$. Observe that f is an isometry from $(\mathbb{R}, d)$ to the subspace $(-1, 1)$

of the real line $\mathbb{R}$. It is easy to observe that $(\mathbb{R}, d)$ is topologically equivalent to the real line but $(\mathbb{R}, d)$ is not complete as $(-1, 1)$ is not complete.

Theorem 5.2.16. *The set of real numbers $\mathbb{R}$ with the usual metric is complete.*

Proof. Let (x_n) be a Cauchy sequence in $\mathbb{R}$. Then it is bounded in $\mathbb{R}$. Let l and u be lower and upper bounds of the set $\{x_n \mid n \in \mathbb{R}\}$, respectively, that is,

$$l \leq x_n \leq u \quad \text{for all } n \in \mathbb{N}.$$

Consider the set

$$A = \{y \in \mathbb{R} \mid y \leq x_m \text{ for infinitely many terms } x_m \text{ of the sequence}\}.$$

Note that $A \neq \emptyset$ since $l \in A$. Also, note that A is bounded above by $u \in \mathbb{R}$. Then by the least upper bound property of $\mathbb{R}$ the supremum of A exists in $\mathbb{R}$. Let $x \in \mathbb{R}$ be the supremum of A. We will show that $x_n \to x$.

Let $\epsilon > 0$. Since x is the supremum of A, there is $z \in A$ such that

$$x - \frac{\epsilon}{2} < z \leq x.$$

Since $z \in A$, infinitely many terms of the sequence are greater than or equal to $x - \frac{\epsilon}{2}$. Note that $x + \frac{\epsilon}{2} \notin A$; otherwise, x would not be an upper bound of A. Therefore at most finitely many terms of the sequence are greater than or equal to $x + \frac{\epsilon}{2}$. Hence the open interval $(x - \frac{\epsilon}{2}, x + \frac{\epsilon}{2})$ contains infinitely many terms of the sequence. Since (x_n) is Cauchy, there is $k \in \mathbb{N}$ such that

$$m, n \geq k \Rightarrow |x_m - x_n| < \frac{\epsilon}{2}.$$

Choose $x_t \in (x - \frac{\epsilon}{2}, x + \frac{\epsilon}{2})$ such that $t > k$. Now for $s \geq k$, we have

$$|x_s - x| \leq |x_s - x_t| + |x_t - x|$$
$$< \frac{\epsilon}{2} + \frac{\epsilon}{2}$$
$$< \epsilon.$$

Thus $x_n \to x$. □

By Example 5.1.10 we can see that $\mathbb{R}^n$ with the usual metric is complete. The completeness is not preserved under a homeomorphism as $\mathbb{R}$ is complete but $(0, 1)$ is not complete. A homeomorphism can send a Cauchy sequence to a non-Cauchy sequence. For example, $(\frac{1}{n+1})$ is a Cauchy sequence in $(0, 1)$, and the map $f : (0, 1) \to (1, \infty)$ defined by $f(x) = \frac{1}{x}$ is a homeomorphism, but $(f(\frac{1}{n+1})) = (n + 1)$ is not a Cauchy sequence in $(1, \infty)$. We can easily observe that the completeness and being a Cauchy sequence are preserved under an isometry. We leave this as an exercise. We can also easily observe the following proposition, whose proof is left as an exercise.

Proposition 5.2.17. *Let d_1 and d_2 be Lipschitz equivalent metrics on a nonempty set X. Then a sequence (x_n) is Cauchy in (X, d_1) if and only if it is Cauchy in (X, d_2). In particular, (X, d_1) is complete if and only if (X, d_2) is complete.*

Topologically equivalent metrics on a metric space X may not preserve the completeness. Let $f : \mathbb{R} \to (0, 1)$ be a homeomorphism. Define the metric ρ on $\mathbb{R}$ by

$$\rho(x, y) = |f(x) - f(y)|.$$

Then f is an isometry from $(\mathbb{R}, \rho)$ to $(0, 1)$ with the usual metric. This shows that $(\mathbb{R}, \rho)$ is not complete. Now the identity map $I = f^{-1} \circ f$ is a homeomorphism from $(\mathbb{R}, \rho)$ to $\mathbb{R}$ with the usual metric. This shows that ρ and the usual metric are topologically equivalent metrics on $\mathbb{R}$.

Theorem 5.2.18. *The normed space $(C_{\mathbb{F}}[a, b], \|\cdot\|_\infty)$ of continuous maps from $[a, b]$ to $\mathbb{F}$, where*

$$\|f\|_\infty = \sup\{|f(x)| \mid x \in [a, b]\},$$

is complete.

Proof. Let (f_n) be a Cauchy sequence in $C_{\mathbb{F}}[a, b]$. Let $\epsilon > 0$. Then there exists $k \in \mathbb{N}$ such that

$$m, n \geq k \Rightarrow \|f_m - f_n\|_\infty < \frac{\epsilon}{3}.$$

This implies that for $m, n \geq k$ and for all $x \in [a, b]$,

$$|f_m(x) - f_n(x)| \leq \|f_m - f_n\|_\infty < \frac{\epsilon}{3}.$$

Therefore, for each $x \in [a, b]$, $(f_n(x))$ is a Cauchy sequence in $\mathbb{F}$. Since $\mathbb{F}$ is complete, $f_n(x) \to y_x$ in $\mathbb{F}$ for all $x \in [a, b]$. This defines the map $f : [a, b] \to \mathbb{F}$ by $f(x) = y_x$. We claim that $f_n \to f$ and $f \in C_{\mathbb{F}}[a, b]$.

Since the absolute value defines a continuous map, by Proposition 5.1.18, for all $n \geq k$ and for all $x \in [a, b]$, we have

$$
\begin{aligned}
|f(x) - f_n(x)| &= \left| \lim_{n \to \infty} f_m(x) - f_n(x) \right| \\
&= \lim_{n \to \infty} |f_m(x) - f_n(x)| \\
&\leq \frac{\epsilon}{3}.
\end{aligned}
$$

Since k is independent of x, $f_n \to f$.

Let $n \geq k$ and $u \in [a, b]$. Since f_n is continuous, there is $\delta > 0$ such that

$$|x - u| < \delta \Rightarrow |f_n(x) - f_n(u)| < \frac{\epsilon}{3}.$$

Let $u \in [a, b]$ be such that $|x - u| < \delta$ and $n \geq k$. Then

$$|f(x) - f(u)| \leq |f(x) - f_n(x)| + |f_n(x) - f_n(u)| + |f_n(u) - f(u)|$$
$$< \frac{\epsilon}{3} + \frac{\epsilon}{3} + \frac{\epsilon}{3}$$
$$= \epsilon.$$

This shows that f is continuous. Hence $f \in C_{\mathbb{F}}[a, b]$. $\qquad\square$

Example 5.2.19. The normed space $C_{\mathbb{F}}[a, b]$ is not complete with the norm defined by

$$\|f\|_1 = \int_a^b |f(x)| dx.$$

Consider the sequence (f_n) defined by

$$f_n(x) = \begin{cases} n(x - a) & \text{if } x \in [a, a + \frac{1}{n}], \\ 1 & \text{if } x \in [a + \frac{1}{n}, b]. \end{cases}$$

Note that each f_n is continuous map by the pasting lemma. For $m > n$, we have

$$\|f_m - f_n\|_1 = (m - n) \int_a^{a + \frac{1}{m}} (x - a) dx + \int_{a + \frac{1}{m}}^{a + \frac{1}{n}} (1 - n(x - a)) dx$$
$$= \frac{1}{2n} - \frac{1}{2m}$$
$$< \frac{1}{2n}.$$

Therefore (f_n) is a Cauchy sequence in $C_{\mathbb{F}}[a, b]$.

For each $x \in [a, b]$, $f_n(x) \to f(x)$, where

$$f(x) = \begin{cases} 0 & \text{if } x = a, \\ 1 & \text{if } x \in (a, b]. \end{cases}$$

If $C_{\mathbb{F}}[a, b]$ is complete, then $f_n \to f$. This is a contradiction since $f \notin C_{\mathbb{F}}[a, b]$.

Theorem 5.2.20. *The normed space ℓ_p, where $1 \leq p \leq \infty$, is complete.*

Proof. We will prove the completeness for $1 \leq p < \infty$. The case $p = \infty$ is left as an exercise.

Let $z \in \ell_p$. Then $z : \mathbb{N} \to \mathbb{F}$ is such that

$$\sum_{i=1}^{\infty} |z(i)|^2 < \infty.$$

Let (x_n) be a Cauchy sequence in ℓ_p. Let $\epsilon > 0$. Then there is $k \in \mathbb{N}$ such that

$$m, n \geq k \Rightarrow \|x_m - x_n\|_p < \epsilon.$$

For $i \in \mathbb{N}$ and $m, n \geq k$, we have

$$|x_m(i) - x_n(i)| \leq \|x_m - x_n\|_p < \epsilon.$$

This implies that $(x_n(i))$ is a Cauchy sequence in $\mathbb{F}$ for each $i \in \mathbb{N}$. Since $\mathbb{F}$ is complete, $x_n(i) \to y_i$ in $\mathbb{F}$. Hence we get the map $x : \mathbb{N} \to \mathbb{F}$ defined by $x(i) = y_i$. Again, since (x_n) is a Cauchy sequence in ℓ_p, there is $k_1 \in \mathbb{N}$ such that

$$m, n \geq k_1 \Rightarrow \sum_{i=1}^{N} |x_m(i) - x_n(i)|^p \leq \|x_m - x_n\|_p^p < \frac{\epsilon^p}{2^p}.$$

Since the absolute value defines a continuous map, we have

$$n \geq k_1 \Rightarrow \sum_{i=1}^{N} |x - x_n(i)|^p \leq \frac{\epsilon^p}{2^p}. \tag{5.3}$$

Since equation (5.3) is true for all $N \in \mathbb{N}$, we have

$$n \geq k_1 \Rightarrow \sum_{i=1}^{\infty} |x - x_n(i)|^p \leq \frac{\epsilon^p}{2^p}.$$

In other words,

$$\|x_n - x\|_p \leq \frac{\epsilon}{2} < \epsilon.$$

This shows that $x_n \to x$. Next, we show that $x \in \ell_p$. Now by the triangle inequality we have

$$\|x\|_p := \left(\sum_{i=1}^{\infty} |x(i)|^p \right)^{\frac{1}{p}}$$

$$\leq \left(\sum_{i=1}^{\infty} |x(i) - x_n(i)|^p \right)^{\frac{1}{p}} + \left(\sum_{i=1}^{\infty} |x_n(i)|^p \right)^{\frac{1}{p}}$$

$$\leq \epsilon^{\frac{1}{p}} + \|x\|_p.$$

This shows that $x \in \ell_p$. $\qquad\qquad\qquad\qquad\qquad\qquad\qquad\qquad\qquad\qquad\quad\square$

Proposition 5.2.21. *A metric space X is complete if and only if every Cauchy sequence in X has a convergent subsequence.*

Proof. Let X be a complete metric space. Then every Cauchy sequence being a subsequence of itself is convergent.

Conversely, suppose every Cauchy sequence in X has a convergent subsequence. Let (x_n) be a Cauchy sequence in X that contains a subsequence x_{n_i} such that $x_{n_i} \to x$ in X. Let $\epsilon > 0$. Then there is $k \in \mathbb{N}$ such that

$$m, n \geq k \Rightarrow d(x_m, x_n) < \frac{\epsilon}{2}.$$

Choose $n_i \in \mathbb{N}$ with $n_i \geq k$ such that

$$d(x_{n_i}, x) < \frac{\epsilon}{2}.$$

Now for $n \geq k$, we have

$$d(x_n, x) \leq d(x_n, x_{n_i}) + d(x_{n_i}, x)$$
$$< \frac{\epsilon}{2} + \frac{\epsilon}{2}$$
$$= \epsilon.$$

Hence $x_n \to x$. $\qquad\qquad\square$

Now we are interested in which sets in a complete metric space are complete as subspaces.

Proposition 5.2.22. *Let X be a complete metric space. Then a set F in X is complete if and only if F is closed.*

Proof. Suppose that a set F in a complete metric space X is complete. Let $x \in X$ be a limit point of F. There is a sequence (x_n) in F such that $x_n \to x$. Since a convergent sequence is Cauchy, (x_n) is Cauchy in F. Since F is complete, $x \in F$. This shows that F is closed in X.

Conversely, suppose that F is a closed set in a complete metric space X. Let (x_n) be a Cauchy sequence in F. Then (x_n) is also a Cauchy sequence in X. Since X is complete, $x_n \to x$ in X. Let

$$A = \{x_n \in F \mid n \in \mathbb{N}\} \subseteq F \subseteq X.$$

Then the set A is either finite or infinite. First, suppose that A is finite. Since (x_n) is convergent in X, the point x has to repeat infinitely times and must be from A. Therefore $x \in F$.

Now suppose that A is infinite. Then x is a limit point of A. Since $A \subseteq F$, x is a limit point of F. Since F is closed, $x \in F$. Hence F is complete. $\qquad\square$

Proposition 5.2.23. *Let W be a finite-dimensional subspace of a normed space V. Then W is closed in V.*

Proof. Since W is finite dimensional over $\mathbb{F}$, we can isometrically identify W with $\mathbb{F}^n$ for some $n \in \mathbb{N}$. Since $\mathbb{F}^n$ is complete, W is complete. Let (x_n) be a sequence in W such that $x_n \to x$ in V. Then (x_n) is Cauchy in W. Since W is complete, $x \in W$. This shows that W is closed. □

We have seen above that a Cauchy sequence in not preserved under a homeomorphism, in particular under a continuous map. We will see below that a Cauchy sequence is preserved under a uniformly continuous map.

Proposition 5.2.24. *Let $f : X \to Y$ be a uniformly continuous map. If (x_n) is a Cauchy sequence in X, then $(f(x_n))$ is a Cauchy sequence in Y.*

Proof. Let (x_n) be a Cauchy sequence in X, and let $y_n = f(x_n)$. Let $\epsilon > 0$. Since f is uniformly continuous, there is $\delta > 0$ such that for all $x, y \in X$, we have

$$d(x,y) < \delta \Rightarrow d(f(x), f(y)) < \epsilon. \qquad (5.4)$$

Since (x_n) is a Cauchy sequence, there is $k \in \mathbb{N}$ such that

$$m, n \geq k \Rightarrow d(x_m, x_n) < \delta. \qquad (5.5)$$

By equations (5.4) and (5.5) we have

$$m, n \geq k \Rightarrow d(f(x_m), f(x_n)) < \epsilon.$$

This shows that $(f(x_n))$ is a Cauchy sequence in Y. □

Theorem 5.2.25 (Continuous extension theorem). *Let A be a subspace of a metric space X, and let $f : A \to Y$ be a uniformly continuous map, where Y is a complete metric space. Then there is a unique continuous map $g : \mathrm{Cl}\, A \to Y$ such that $g(a) = f(a)$ for all $a \in A$.*

Proof. Let $x \in \mathrm{Cl}\, A$. Then there is a sequence (x_n) in A such that $x_n \to x$. Note (x_n) is a Cauchy sequence in A. By Proposition 5.2.24 $(f(x_n))$ is a Cauchy sequence in Y. Since Y is complete, $f(x_n) \to y_x$ in Y.

Suppose (z_n) is another sequence in A such that $z_n \to x$. Then by a similar argument, $f(z_n) \to z_x$ in Y. Let $\epsilon > 0$. Since f is uniformly continuous on A, there is $\delta > 0$ such that for all $a, b \in A$, we have

$$d(a,b) < \delta \Rightarrow d(f(a), f(b)) < \frac{\epsilon}{3}. \qquad (5.6)$$

Since $x_n \to x$, there is $k_1 \in \mathbb{N}$ such that

$$n \geq k_1 \Rightarrow d(x_n, x) < \frac{\delta}{2}. \qquad (5.7)$$

Similarly, as $z_n \to x$, there is $k_2 \in \mathbb{N}$ such that

$$n \geq k_2 \Rightarrow d(z_n, x) < \frac{\delta}{2}. \tag{5.8}$$

Let $k' = \max\{k_1, k_2\}$. Then for $k' \in \mathbb{N}$, equations (5.7) and (5.8) hold. Then

$$n \geq k' \Rightarrow d(x_n, z_n) \leq d(x_n, x) + d(z_n, x)$$
$$< \frac{\delta}{2} + \frac{\delta}{2}$$
$$= \delta. \tag{5.9}$$

This implies that $d(x_n, z_n) \to 0$. By equations (5.6) and (5.8) we have

$$n \geq k' \Rightarrow d(f(x_n), f(z_n)) < \frac{\epsilon}{3}. \tag{5.10}$$

Since $f(x_n) \to y_x$, there is $k_3 \in \mathbb{N}$ such that

$$n \geq k_3 \Rightarrow d(f(x_n), y_x) < \frac{\epsilon}{3}. \tag{5.11}$$

Since $f(z_n) \to z_x$, there is $k_4 \in \mathbb{N}$ such that

$$n \geq k_3 \Rightarrow d(f(z_n), z_x) < \frac{\epsilon}{3}. \tag{5.12}$$

Let $k = \max\{k', k_3, k_4\}$. Then for $k \in \mathbb{N}$, equations (5.10), (5.11), and (5.12) hold. Now for $n \geq k$, we have

$$d(y_x, z_x) \leq d(y_x, f(x_n)) + d(f(x_n), f(z_n)) + d(f(z_n), z_x)$$
$$< \frac{\epsilon}{3} + \frac{\epsilon}{3} + \frac{\epsilon}{3}$$
$$= \epsilon.$$

Since $\epsilon > 0$ is arbitrary, $y_x = z_x$. Thus we have the map $g : \mathrm{Cl}\, A \to Y$ defined by

$$g(x) = \begin{cases} f(x) & \text{if } x \in A, \\ \lim_{n \to \infty} f(x_n) & \text{if } x \in \mathrm{Cl}\, A \setminus A, \end{cases}$$

where (x_n) is a sequence in A such that $x_n \to x$.

Now we will show that g is continuous. We will in fact show that g is uniformly continuous on $\mathrm{Cl}\, A$. Since f is uniformly continuous on A, for a given $\epsilon > 0$, there is $\delta > 0$ such that for $a, b \in A$, we have

$$d(a, b) < \delta \Rightarrow d(f(a), f(b)) < \epsilon. \tag{5.13}$$

Let $x, y \in \mathrm{Cl}\, A$ be such that $d(x, y) < \frac{\delta}{3}$. Choose sequences (x_n) and (y_n) in A such that $x_n \to x$ and $y_n \to y$. Then there are $k_5, k_6 \in \mathbb{N}$ such that

$$n \geq k_5 \Rightarrow d(x_n, x) < \frac{\delta}{3} \tag{5.14}$$

and

$$n \geq k_6 \Rightarrow d(y_n, y) < \frac{\delta}{3}. \tag{5.15}$$

Let $k'' = \max\{k_5, k_6\}$. Then for $k'' \in \mathbb{N}$, equations (5.14) and (5.14) hold. Now, for $n \geq k''$, we have

$$
\begin{aligned}
d(x_n, y_n) &\leq d(x_n, x) + d(x, y) + d(y_n, y) \\
&< \frac{\delta}{3} + \frac{\delta}{3} + \frac{\delta}{3} \\
&= \delta.
\end{aligned}
$$

Now for $n \geq k''$, we have

$$
\begin{aligned}
d(g(x), g(y)) &= d\left(\lim_{n \to \infty} f(x_n), \lim_{n \to \infty} f(y_n) \right) \\
&= \lim_{n \to \infty} d(f(x_n), f(y_n)) \\
&\leq \epsilon.
\end{aligned}
$$

This shows that g is uniformly continuous on $\mathrm{Cl}\, A$.

Now we will prove that such a map g is unique. Let $h : \mathrm{Cl}\, A \to Y$ be a continuous map such that $h(a) = f(a)$ for all $a \in A$. If $x \in A$, then $g(x) = f(x) = h(x)$. Let $x \in \mathrm{Cl}\, A \setminus A$. Let (x_n) be a sequence in A such that $x_n \to x$. Then

$$
\begin{aligned}
g(x) &= \lim_{n \to \infty} f(x_n) \\
&= \lim_{n \to \infty} h(x_n) \\
&= h\left(\lim_{n \to \infty} x_n \right) \\
&= h(x).
\end{aligned}
$$

This shows that $g = h$. □

Example 5.2.26. The map $f : (0, 1) \to \mathbb{R}$ defined by $f(x) = \frac{1}{x}$ is not uniformly continuous on $(0, 1)$; otherwise, it would have a continuous extension g of f on $[0, 1]$, which is not possible.

Let us observe an important property of a complete metric space.

Theorem 5.2.27 (Cantor's intersection theorem). *Let X be a complete metric space, and let (F_n) be a sequence of closed, bounded, and nonempty sets in X such that*
(i) $F_n \supseteq F_{n+1}$ *for all $n \in \mathbb{N}$ and*
(ii) $\operatorname{diam}(F_n) \to 0$.

Then $\bigcap_{n \in \mathbb{N}} F_n$ is a singleton set.

Proof. For each $n \in \mathbb{N}$, choose $x_n \in F_n$. Let $\epsilon > 0$. Since $\operatorname{diam}(F_n) \to 0$, there is $k \in \mathbb{N}$ such that

$$n \geq k \Rightarrow \operatorname{diam}(F_n) < \epsilon.$$

In particular, $\operatorname{diam}(F_k) < \epsilon$. This implies that

$$m, n \geq k \Rightarrow d(x_m, x_n) < \epsilon.$$

Therefore (x_n) is a Cauchy sequence in X. Since X is complete, $x_n \to x$ in X. We claim that $\bigcap_{n \in \mathbb{N}} F_n = \{x\}$.

We first observe that x lies in each F_n. First, suppose that the set of the terms of the sequence (x_n) is finite. Then the sequence (x_n) is constant on the tail. Clearly, $x \in F_n$ in this case. Now suppose that the set of the terms of the sequence (x_n) is infinite. Let $E_n = \{x_l \mid l \geq n\}$. Then x is a limit point of E_n for all $n \in \mathbb{N}$. Since $E_n \subseteq F_n$, x is a limit point of F_n. Since F_n is closed, $x \in F_n$ for all $n \in \mathbb{N}$.

Let $y \in \bigcap_{n \in \mathbb{N}} F_n$ be such that $y \neq x$. Then we can note that $\operatorname{diam}(F_n) \nrightarrow 0$. This is a contradiction. $\qquad\square$

Let us see an application of Cantor's intersection theorem. Let b be a positive integer such that $b \geq 2$. Note that any positive real number x can be represented as $n + r$, where $n \in \mathbb{Z}$ and $r \in [0, 1]$. Applying the division algorithm, we can show that any positive integer can be represented in the base b. Now we will show that any real number in $[0, 1]$ can also be represented in the base b. We will adopt the following steps.

In the first step, divide $[0, 1]$ into b equal parts as follows:

$$\frac{0}{b} < \frac{1}{b} < \cdots < \frac{a}{b} < \frac{a+1}{b} < \cdots < \frac{b-1}{b} < \frac{b}{b}.$$

If $[\frac{a}{b}, \frac{a+1}{b}]$ is a subinterval from the first step, then $a \in \{0, 1, \ldots, b-1\}$. By the notation $(\frac{a_1 a_2 \cdots a_k}{b^k})$ we mean that

$$\frac{a_1}{b} + \frac{a_1}{b^2} + \cdots + \frac{a_k}{b^k}.$$

Now divide the first step subintervals into b equal parts. Let $A_1(a_1) = [\frac{a_1}{b}, \frac{a_1+1}{b}]$ be one of the subintervals. Then we have

$$\left(\frac{a_1 0}{b^2}\right) < \left(\frac{a_1 1}{b^2}\right) < \cdots < \left(\frac{a_1 a_2}{b^2}\right) < \left(\frac{a_1 a_2 + 1}{b^2}\right) < \cdots < \frac{a_1 + 1}{b}.$$

In this step, we get the subintervals $A_2(a_1, a_2) = [(\frac{a_1 a_2}{b^2}), (\frac{a_1 a_2 + 1}{b^2})]$, where $a_2 \in \{0, 1, \ldots, b - 1\}$. Inductively, in the nth step, we get the subinterval

$$A_n(a_1, \ldots, a_n) = \left[\left(\frac{a_1 a_2 \cdots a_n}{b^n}\right), \left(\frac{a_1 a_2 \cdots a_n + 1}{b^n}\right)\right],$$

where $a_n \in \{0, 1, \ldots, b - 1\}$. Thus we get the decreasing sequence of closed intervals

$$[0, 1] \supseteq A_1(a_1) \supseteq \cdots \supseteq A_n(a_1, \ldots, a_n) \supseteq \cdots.$$

By Cantor's intersection theorem, $\bigcap_{n=1}^{\infty} A_n(a_1, \ldots, a_n)$ is a singleton set, say $\{r\}$. Note that for any $n \in \mathbb{N}$, we have

$$\frac{a_1}{b} + \frac{a_1}{b^2} + \cdots + \frac{a_n}{b^n} \leq r < \frac{a_1}{b} + \frac{a_1}{b^2} + \cdots + \frac{a_n}{b^n} + \frac{1}{b^n}.$$

This implies that

$$\left| r - \frac{a_1}{b} + \frac{a_1}{b^2} + \cdots + \frac{a_n}{b^n} \right| < \frac{1}{b^n}.$$

This shows that the series

$$\frac{a_1}{b} + \frac{a_1}{b^2} + \cdots + \frac{a_n}{b^n} + \cdots$$

converges to r. We will represent r as $(0.a_1 a_2 \cdots)_b$ in the base b.

Now we claim that each real number in $[0, 1]$ can be represented in base b. Let $r \in [0, 1]$. Note that r belongs to at least one subinterval at each step of division of the intervals into b equal parts. Therefore r is the intersection point of one sequence of decreasing intervals

$$[0, 1] \supseteq A_1(a_1) \supseteq \cdots \supseteq A_n(a_1, \ldots, a_n) \supseteq \cdots.$$

Consider the end point $(\frac{a_1 a_2 \cdots a_n}{b^n})$ of the subinterval at any particular step. We wish to find the representation of such an end point in the base b. We can further divide the subintervals either considering this point on the left of each subintervals afterward or considering this point on the right of each subintervals afterward. Thus such points have two representations; one terminates in the digit zero, and the other in $b - 1$. The points other than end points lie in the unique subintervals at each step. Therefore the other points have unique representations in the base b.

5.3 Completion of a metric space

In this section, we will complete a metric space. We will show that it satisfies a universal property.

Definition 5.3.1. A metric space $(\hat{X}, \hat{d})$ is called a completion of a metric space (X, d) if $(\hat{X}, \hat{d})$ is complete and (X, d) is isometrically embedded in $(\hat{X}, \hat{d})$ as a dense set.

First, we will prove that a completion (if exists) of a metric satisfies some universal property. Later, we will show that the completion of each metric space exists.

Theorem 5.3.2. *Let (X, d_X) be a metric space, and let $(\hat{X}, \hat{d})$ be a complete metric space. Let $f : X \to \hat{X}$ be an isometric embedding. Then the image $f(X)$ is dense in $\hat{X}$ if and only if given any complete metric space (Y, d_Y) and an isometric embedding $g : X \to Y$, there is a unique isometry $\psi : \hat{X} \to Y$ such that the following diagram is commutative:*

$$
\begin{array}{ccc}
X & \xrightarrow{\;f\;} & \hat{X} \\
 & \searrow{\scriptstyle g} & \downarrow{\scriptstyle \psi} \\
 & & Y
\end{array}
$$

Proof. Suppose that $f(X)$ is dense in $\hat{X}$. Let (Y, d_Y) be a complete metric space, and let g be an isometric embedding from X to Y. Note that $f^{-1} : f(X) \to X$ is an isometry. Then $h = g \circ f^{-1}$ is an isometric embedding from $f(X)$ to Y. Define the map $\psi : \hat{X} \to Y$ by

$$
\psi(x) = \begin{cases} h(x) & \text{if } x \in f(X), \\ \lim_{n \to \infty} h(x_n) & \text{if } x \in \hat{X} \setminus f(X), \end{cases}
$$

where (x_n) is a sequence in $f(X)$ such that $x_n \to x$. We can indeed check that ψ is well defined and ψ is an isometry. We can also check that such a map ψ is unique. Now for all $x \in X$, we have

$$
\begin{aligned}
\psi \circ f(x) &= \psi(f(x)) \\
&= h(f(x)) \\
&= g \circ f^{-1}(f(x)) \\
&= g(x).
\end{aligned}
$$

This implies that $\psi \circ f = g$.

If $\phi : \hat{X} \to Y$ is an isometry making the above diagram commutative, then

$$
\phi \circ f = g = \psi \circ f.
$$

Since $f(X)$ is dense in $\hat{X}$, $\psi = \phi$.

Now suppose the converse assumption holds. Since a closed set in a complete metric space is complete, $\mathrm{Cl}f(X)$ is complete. Note that $f : X \to \mathrm{Cl}f(X)$ is complete. By the assumption there is a unique isometry $\psi : \hat{X} \to \mathrm{Cl}f(X)$ such that $\psi \circ f = f$. This implies that ψ is the identity map on $f(X)$. Therefore ψ is the identity map on $\mathrm{Cl}f(X)$. This shows that $\mathrm{Cl}f(X) = \hat{X}$. Hence $f(X)$ is dense in $\hat{X}$. $\qquad\square$

Corollary 5.3.3. *Suppose that a completion of a metric space (X, d) exists. Then it is unique up to isometry.*

Proof. Let (X_1, d_1) and (X_2, d_2) be two completions of a metric space (X, d). Let $i_1 : X \to X_1$ and $i_2 : X \to X_2$ be isometric embeddings such that $i_1(X)$ and $i_2(X)$ are dense in X_1 and X_2, respectively. By Theorem 5.3.2 there exists a unique isometry $\psi : X_1 \to X_2$ such that $\psi \circ i_1 = i_2$. $\qquad\square$

Now we show that the completion of each metric space exists. First, we observe the following:

Proposition 5.3.4. *Let (X, d) be a metric space, and let A be a dense set in X. Suppose that any Cauchy sequence (x_n) in A converges in X. Then X is complete.*

Proof. Let (x_n) be a Cauchy sequence in X. Let $\epsilon > 0$. Then there is $N \in \mathbb{N}$ such that

$$r, s \geq N \Rightarrow d(x_r, x_s) < \frac{\epsilon}{3}.$$

Since A is dense in X, for each x_n, there is a sequence (x_n^k) in A such that $x_n^k \to x_n$ as $k \to \infty$. Note that (x_n^k) is a Cauchy sequence in A. Thus there is $k_n \in \mathbb{N}$ such that

$$m \geq k_n \Rightarrow d(x_n^m, x_n) < \frac{\epsilon}{3}.$$

In particular, we have

$$d(x_n^{k_n}, x_n) < \frac{\epsilon}{3}.$$

We can choose $k_n \in \mathbb{N}$ such that

$$k_1 < k_2 < \cdots.$$

Next, we show that $(x_n^{k_n})_{n \in \mathbb{N}}$ is a Cauchy sequence in A. Now for $r, s \geq N$, we have

$$d(x_r^{k_r}, x_s^{k_s}) \leq d(x_r^{k_r}, x_r) + d(x_r, x_s) + d(x_s, x_s^{k_s})$$
$$< \frac{\epsilon}{3} + \frac{\epsilon}{3} + \frac{\epsilon}{3}$$
$$= \epsilon.$$

Hence $(x_n^{k_n})$ is a Cauchy sequence in A. By the assumption there is $x \in X$ such that $x_n^{k_n} \to x$. Then for arbitrary $\epsilon > 0$, there is $M \in \mathbb{N}$ such that

$$n \geq M \Rightarrow d(x_n^{k_n}, x) < \frac{\epsilon}{2}.$$

Finally, we show that $x_n \to x$. For $n \geq M$, we have

$$d(x_n, x) \leq d(x_n, x_n^{k_n}) + d(x_n^{k_n}, x)$$
$$< \frac{\epsilon}{2} + \frac{\epsilon}{2}$$
$$= \epsilon.$$

This shows that X is complete. $\qquad\qquad\qquad\qquad\qquad\qquad\qquad\qquad\qquad\qquad \square$

Theorem 5.3.5. *Every metric space has a completion.*

Proof. Let (X, d) be a metric space. Let Z be the collection of all Cauchy sequences in X. Let us define a relation R on Z as

$$((x_n), (y_n)) \in R \quad \text{if and only if} \quad \lim_{n \to \infty} d(x_n, y_n) = 0.$$

We can easily show that R is an equivalence relation on Z. Let $\hat{X}$ denote the quotient set

$$Z/R = \{\overline{(x_n)} \mid (x_n) \in Z\}.$$

Let (x_n) and (y_n) be Cauchy sequences in X. Then we can choose $k \in \mathbb{N}$ such that

$$m, n \geq k \Rightarrow d(x_m, x_n) < \frac{\epsilon}{2}$$

and

$$d(y_m, y_n) < \frac{\epsilon}{2}.$$

Now by the inequality

$$|d(a, b) - d(c, d)| \leq d(a, c) + d(b, d)$$

we see that for $m, n \geq k$,

$$|d(x_m, y_m) - d(x_n, y_n)| \leq d(x_m, x_n) + d(y_m, y_n)$$
$$< \frac{\epsilon}{2} + \frac{\epsilon}{2}$$
$$= \epsilon.$$

Therefore $(d(x_n, y_n))$ is a Cauchy sequence in $\mathbb{R}$. Since $\mathbb{R}$ is complete, $\lim_{n \to \infty} d(x_n, y_n)$ exists.

Now let (a_n) and (b_n) be two Cauchy sequences such that

$$\overline{(a_n)} = \overline{(x_n)} \quad \text{and} \quad \overline{(b_n)} = \overline{(y_n)}.$$

Note that

$$d(x_n, y_n) \le d(x_n, a_n) + d(a_n, b_n) + d(b_n, y_n).$$

This implies that

$$\lim_{n \to \infty} d(x_n, y_n) \le \lim_{n \to \infty} d(a_n, b_n).$$

By a similar argument we get

$$\lim_{n \to \infty} d(a_n, b_n) \le \lim_{n \to \infty} d(x_n, y_n).$$

Hence

$$\lim_{n \to \infty} d(a_n, b_n) = \lim_{n \to \infty} d(x_n, y_n).$$

Thus the above argument shows that we have the map

$$\hat{d} : \hat{X} \times \hat{X} \to \mathbb{R}$$

defined by

$$\hat{d}(\hat{x}, \hat{y}) = \lim_{n \to \infty} d(x_n, y_n),$$

where $\hat{x} = \overline{(x_n)}$ and $\hat{y} = \overline{(y_n)}$. We claim that $\hat{d}$ is a metric on $\hat{X}$. Clearly, $\hat{d}(\hat{x}, \hat{y}) \ge 0$ and $\hat{d}(\hat{x}, \hat{y}) = \hat{d}(\hat{y}, \hat{x})$. Also,

$$\hat{d}(\hat{x}, \hat{y}) = 0 \iff \lim_{n \to \infty} d(x_n, y_n) = 0$$
$$\iff ((x_n), (y_n)) \in R$$
$$\iff \hat{x} = \hat{y}.$$

Let $\hat{x}, \hat{y}, \hat{z} \in \hat{X}$. Then

$$\hat{d}(\hat{x}, \hat{y}) + \hat{d}(\hat{y}, \hat{z}) = \lim_{n \to \infty} d(x_n, y_n) + \lim_{n \to \infty} d(y_n, z_n)$$
$$\ge \lim_{n \to \infty} d(x_n, z_n)$$
$$= \hat{d}(\hat{x}, \hat{z}).$$

Therefore $\hat{d}$ is a metric on $\hat{X}$. Now we prove that X is isometrically embedded in $\hat{X}$. Let $x \in X$. Then the constant sequence (x) is a Cauchy sequence in X. Let Y be the set of all equivalence classes of constant sequences in X. Let (x) and (y) be distinct constant sequences in X. Then $x \neq y$. Now

$$\hat{d}(\overline{(x)}, \overline{(y)}) = \lim_{n \to \infty} d(x, y) = d(x, y) \neq 0.$$

This means that (x) and (y) are in the distinct equivalence classes. Hence each element of Y contains only one constant sequence. Let us define the map

$$f : X \to Y \subseteq \hat{X}$$

by $f(x) = \overline{(x)}$. Clearly, $\mathrm{Im} f(X) = Y$. Now

$$\hat{d}(f(x), f(y)) = \hat{d}(\overline{(x)}, \overline{(y)})$$
$$= \lim_{n \to \infty} d(x, y)$$
$$= d(x, y).$$

Therefore X is isometrically embedded in $\hat{X}$.

Next, we will show that $Y = f(X)$ is dense in $\hat{X}$. Let $\hat{x} \in \hat{X}$ and $\epsilon > 0$. Suppose that (x_n) is a Cauchy sequence in $\hat{X}$. Then there is $k \in \mathbb{N}$ such that

$$m, n \geq k \implies d(x_m, x_n) < \frac{\epsilon}{2}.$$

In particular, we have

$$m \geq k \implies d(x_m, x_k) < \frac{\epsilon}{2}.$$

Let $z = x_k$. Consider the constant sequence (z). Let $\hat{z} = \overline{(z)}$. Then

$$\hat{d}(\hat{x}, \hat{z}) = \lim_{m \to \infty} d(x_m, z)$$
$$= \lim_{m \to \infty} d(x_m, x_k)$$
$$\leq \frac{\epsilon}{2}$$
$$< \epsilon.$$

This implies that

$$\hat{z} \in B(\hat{x}, \epsilon) \cap Y.$$

Since $\epsilon > 0$ is arbitrary, $\mathrm{Cl}\, Y = \hat{X}$. This shows that Y is dense in $\hat{X}$.

Finally, we show that $\hat{X}$ is complete. Let $\hat{x}_n$ be a Cauchy sequence in $f(X) \subseteq \hat{X}$. Let each $\hat{x}_n$ be represented by the constant sequence

$$(x^n, x^n, \ldots).$$

Equivalently, we have $f(x^n) = \hat{x}_n$. Let $\epsilon > 0$. Then there is $k \in \mathbb{N}$ such that

$$m, n \geq k \Rightarrow \hat{d}(\hat{x}_m, \hat{x}_n) < \epsilon.$$

Now for $m, n \geq k$, we have

$$\begin{aligned} d(x^m, x^n) &= \hat{d}(f(x^m), f(x^n)) \\ &= \hat{d}(\hat{x}_m, \hat{x}_n) \\ &< \epsilon. \end{aligned}$$

This implies that $(x^1, x^2, \ldots)$ is a Cauchy sequence in X. We can similarly show the reverse implication that if $(x^1, x^2, \ldots)$ is a Cauchy sequence in X, then $f(x^n) = \hat{x}_n$ is a Cauchy sequence in $\hat{X}$.

Let $\hat{x}$ denote the equivalence class of the sequence

$$(x^1, x^2, \ldots)$$

in $\hat{X}$. We will show that $\hat{x}_n \to \hat{x}$. Let $\epsilon > 0$. Since $(x^1, x^2, \ldots)$ is a Cauchy sequence in X, there $k \in \mathbb{N}$ such that

$$m, n \geq k \Rightarrow d(x^m, x^n) < \frac{\epsilon}{2}.$$

Therefore for each $m \geq k$, we have

$$\hat{d}(\hat{x}_m, \hat{x}) = \lim_{n \to \infty} d(x^m, x^n) \leq \frac{\epsilon}{2} < \epsilon.$$

Therefore by Proposition 5.3.4 $\hat{X}$ is complete. □

We have shown the existence of a completion of a metric space through a construction. Alternatively, we can show the existence as follows.

Let (X, d_X) and (Y, d_y) be metric spaces. Let $B(X, Y)$ denote the set of all bounded maps from X to Y. Let $f, g \in B(X, Y)$. We can check that

$$d_\infty(f, g) = \sup\{d_Y(f(x), g(x)) \mid x \in X\}$$

defines a metric on $B(X, Y)$.

Proposition 5.3.6. *Let (X, d_X) and (Y, d_Y) be metric spaces. Then (Y, d_Y) is complete if and only if $(B(X, Y), d_\infty)$ is complete.*

Proof. Suppose that (Y, d_Y) is complete. Let (f_n) be a Cauchy sequence in $B(X, Y)$. Let $\epsilon > 0$. Then there is $k \in \mathbb{N}$ such that

$$m, n \geq k \Rightarrow d_\infty(f_m, f_n) < \epsilon.$$

Note that $d_Y(f_m(x), f_n(x)) \leq d_\infty(f_m, f_n)$ for $x \in X$. Then for $m, n \geq k$ and $x \in X$, we have

$$d_Y(f_m(x), f_n(x)) < \epsilon. \tag{5.16}$$

This implies that $(f_n(x))$ is a Cauchy sequence in Y for each $x \in X$. Since Y is complete, $f_n(x) \to y_x$ in Y. This defines the map $g : X \to Y$ by $g(x) = y_x$. Suppose that $X \times X$ is provided a metric such that the metric d is continuous. Let us take the limit as $m \to \infty$ in equation (5.16). By the continuity of d we have

$$d_Y(f(x), f_n(x)) = d_Y\left(\lim_{m \to \infty} f_m(x), f_n(x)\right)$$
$$= \lim_{m \to \infty} d_Y(f_m(x), f_n(x))$$
$$\leq \epsilon.$$

This implies that $d_\infty(f_n, f) \leq \epsilon$. This shows that $f \in B(X, Y)$ and $f_n \to f$. Hence $B(X, Y)$ is complete.

Conversely, suppose that $B(X, Y)$ is complete. Let (y_n) be a Cauchy sequence in Y. For each $n \in \mathbb{N}$, let $f_n : X \to Y$ be the constant map $f_n(x) = y_n$ for all $x \in X$. Note that each f_n is a bounded map and $d_\infty(f_m, f_n) = d_Y(y_m, y_n)$. Therefore (f_n) is a Cauchy sequence in $B(X, Y)$. Since $B(X, Y)$ is complete, $f_n \to f$ in $B(X, Y)$.

Let $\epsilon > 0$. Then there is $k \in \mathbb{N}$ such that

$$n \geq k \Rightarrow d_\infty(f_n, f) < \frac{\epsilon}{2}.$$

This implies that for each $x \in X$, we have

$$n \geq k \Rightarrow d_Y(f_n(x), f(x)) < \frac{\epsilon}{2}.$$

Now for any $x, y \in X$, we have

$$d_Y(f(x), f(y)) \leq d_Y(f(x), f_n(x)) + d_Y(f_n(x), f_n(y)) + d_Y(f_n(y), f(y))$$
$$= d_Y(f(x), f_n(x)) + d_Y(f_n(y), f(y))$$
$$\leq 2d_\infty(f_n, f)$$
$$< \epsilon.$$

Since $\epsilon > 0$ is arbitrary, $d_Y(f(x), f(y)) = 0$. Therefore $f(x) = f(y)$ for all $x, y \in X$. Hence f is a constant map. Suppose $f(x) = c$ for all $x \in X$. Since $d_Y(y_n, c) = d_\infty(f_n, f)$, $y_n \to c$. This shows that Y is complete. $\qquad \square$

Theorem 5.3.7. *Every metric space has a completion.*

Proof. Let (X, d) be a metric space. Let us fix $c \in X$. For each $a \in X$, define the map $f_a : X \to \mathbb{R}$ by

$$f_a(x) = d(x, a) - d(x, c).$$

Now

$$|f_a(x)| = |d(x, a) - d(x, c)| \le d(a, c).$$

This implies that $f_a \in B(X, \mathbb{R})$, which defines the map $f : X \to B(X, \mathbb{R})$ by $f(a) = f_a$. Let $a, b \in X$. Then

$$
\begin{aligned}
d_\infty(f(a), f(b)) &= d_\infty(f_a, f_b) \\
&= \sup\{|f_a(x) - f_b(x)| \mid x \in X\} \\
&= \sup\{|d(x, a) - d(x, b)| \mid x \in X\} \\
&\le \sup\{d(a, b) \mid x \in X\} \\
&= d(a, b).
\end{aligned}
$$

For a particular value $x = a$, we have

$$d(a, b) = |d(a, a) - d(a, b)| \le \sup\{|d(x, a) - d(x, b)| \mid x \in X\}.$$

This implies that $d_\infty(f(a), f(b)) = d(a, b)$. This shows that f is an isometric embedding of X in $B(X, \mathbb{R})$. Note that $B(X, \mathbb{R})$ is complete as $\mathbb{R}$ is complete. Since $\mathrm{Cl} f(X)$ is a closed set in $B(X, \mathbb{R})$, $\mathrm{Cl} f(X)$ is complete. We can observe that $\mathrm{Cl} f(X)$ is a completion of $f(X)$. Since X is isometric to $f(X)$, $\mathrm{Cl} f(X)$ is a completion of X. $\qquad \square$

Example 5.3.8. In Example 5.2.19, we observed that the space $C[a, b]$ of continuous maps is not complete with respect to the norm

$$\|f\|_1 = \int_a^b |f(x)| dx.$$

Its completion is denoted by $L^1[a, b]$ and is called the space of Lebesgue-integrable functions.

5.4 Completion of $\mathbb{Q}$

In this section, we show that a completion of the set $\mathbb{Q}$ of rational numbers with the usual metric is the real line $\mathbb{R}$ using the construction described in the previous section.

Let Z denote the set of all Cauchy sequences in $\mathbb{Q}$, and let R be the equivalence relation on Z described in the previous section. Let $\hat{\mathbb{Q}}$ denote the quotient set Z/R. Let $\hat{x} = \overline{(x_n)}$ and $\hat{y} = \overline{(y_n)}$ in $\hat{\mathbb{Q}}$. Let us define $\hat{x} + \hat{y}$ and $\hat{x}\hat{y}$ as the equivalence class of $(x_n + y_n)$ and $(x_n y_n)$, respectively. We claim that these are well defined. For this, let $\overline{(x_n)} = \overline{(a_n)}$ and $\overline{(y_n)} = \overline{(b_n)}$. Then

$$(x_n + y_n) - (a_n + b_n) = (x_n - a_n) + (y_n - b_n) \to 0.$$

Therefore $\overline{x_n + y_n} = \overline{a_n + b_n}$. Also, since a Cauchy sequence is bounded, there are real numbers $M_1 > 0$ and $M_2 > 0$ such that $|y_n| < M_1$ and $|a_n| < M_2$. Then

$$\begin{aligned}
|x_n y_n - a_n b_n| &= |x_n y_n - a_n y_n + a_n y_n - a_n b_n| \\
&\leq |x_n - a_n||y_n| + |a_n||y_n - b_n| \\
&\leq |x_n - a_n|M_1 + M_2|y_n - b_n|.
\end{aligned}$$

Hence $x_n y_n - a_n b_n \to 0$. Thus the addition and multiplication are well-defined binary operations on $\hat{\mathbb{Q}}$.

Theorem 5.4.1. *The triple $(\hat{\mathbb{Q}}, +, \cdot)$ is a field, where $+$ and $\cdot$ are binary operations $\hat{\mathbb{Q}}$ defined above.*

Proof. Observe that the equivalence class of the constant sequence (0) and the equivalence class of the constant sequence (1) are the additive and multiplicative identities of $\hat{\mathbb{Q}}$, respectively. We denote these by $\hat{0}$ and $\hat{1}$, respectively. Also, for a given $\hat{x} = \overline{(x_n)}$ in $\hat{\mathbb{Q}}$, $-\hat{x}$ is the equivalence class of $(-x_n)$. We will only prove that if $\hat{x} \in \hat{\mathbb{Q}}$ with $\hat{x} \neq \hat{0}$, then there is $\hat{y} \in \hat{\mathbb{Q}}$ such that $\hat{x}\hat{y} = \hat{1}$. The other conditions of a field are left as exercises.

Let $\hat{x} = \overline{(x_n)} \neq \hat{0}$. This implies that the Cauchy sequence (x_n) does not converge to 0. Since (x_n) is a Cauchy sequence, for a given $\epsilon > 0$, there is $k \in \mathbb{N}$ such that for all $n, m \geq k$, we have $|x_n - x_m| < \epsilon$. In particular, we have

$$n \geq k \Rightarrow |x_n - x_k| < \epsilon.$$

Since $x_n \not\to 0$, we can choose $x_k \neq 0$. Equivalently, we can write

$$n \geq k \Rightarrow x_k - \epsilon < x_n < x_k + \epsilon.$$

We can choose a suitable $\epsilon > 0$ to ensure that

$$n \geq k \Rightarrow x_n \neq 0.$$

Define the sequence (y_n) by $y_n = \frac{1}{x_n}$ for $n \geq k$, letting y_n be any fixed rational number for $1 < n < k$. We can check that (y_n) is a Cauchy sequence. Let $\hat{y} = \overline{(y_n)}$. Then $\hat{x}\hat{y}$ is the equivalence class of the sequence that is constant 1 on the tail. This shows that $\hat{x}\hat{y} = \hat{1}$. $\square$

Let $\hat{x} = \overline{(x_n)} \in \hat{\mathbb{Q}}$. We say that $\hat{x}$ is positive (denoted by $\hat{x} > \hat{0}$) if there are a positive rational number δ and $k \in \mathbb{N}$ such that

$$n \geq k \Rightarrow x_n > \delta.$$

We define the relation $<$ on $\hat{\mathbb{Q}}$ as $\hat{x} = \overline{(x_n)} < \hat{y} = \overline{(y_n)}$ if $\hat{y} - \hat{x} > \hat{0}$. Equivalently, if $\hat{x} < \hat{y}$, then there are a positive rational number δ and $k \in \mathbb{N}$ such that

$$n \geq k \Rightarrow x_n + \delta < y_n.$$

We first observe that $<$ is well defined on $\hat{\mathbb{Q}}$. For this, let $\hat{x} < \hat{y}$. Then there are a positive rational number δ and $k_1 \in \mathbb{N}$ such that

$$n \geq k_1 \Rightarrow x_n + \delta < y_n. \tag{5.17}$$

Suppose that $\overline{(x_n)} = \overline{(a_n)}$ and $\overline{(y_n)} = \overline{(b_n)}$. This implies that

$$\lim_{n \to \infty} |x_n - a_n| = 0 \quad \text{and} \quad \lim_{n \to \infty} |y_n - b_n| = 0.$$

Then there are $k_2, k_3 \in \mathbb{N}$ such that

$$n \geq k_2 \Rightarrow |a_n - x_n| < \frac{\delta}{3} \tag{5.18}$$

and

$$n \geq k_3 \Rightarrow |b_n - y_n| < \frac{\delta}{3}. \tag{5.19}$$

Let $k = \max\{k_1, k_2, k_3\}$. For $k \in \mathbb{N}$, equations (5.17), (5.18), and (5.19) are satisfied. Now for $n \geq k$, we have

$$a_n + \frac{\delta}{3} < x_n + \frac{2\delta}{3} < y_n - \frac{\delta}{3} < b_n.$$

This shows that $\overline{(a_n)} < \overline{(b_n)}$. We now prove that given $\hat{x}, \hat{y} \in \hat{\mathbb{Q}}$, either $\hat{x} < \hat{y}$ or $\hat{x} = \hat{y}$ or $\hat{y} < \hat{x}$.

Let $\hat{x} \neq \hat{y}$. This implies that $|x_n - y_n| \nrightarrow 0$. Therefore there is a positive rational number δ such that for all $k \in \mathbb{N}$, there is $n \geq k$ such that $|x_n - y_n| \geq \delta$.

Since (x_n) and (y_n) are Cauchy sequences, there are $l_1, l_2 \in \mathbb{N}$ such that

$$n \geq l_1 \Rightarrow |x_m - x_n| < \frac{\delta}{3} \tag{5.20}$$

and

$$n \geq l_2 \Rightarrow |y_m - y_n| < \frac{\delta}{3}. \tag{5.21}$$

Let $l' = \max\{l_1, l_2\}$. For l', equations (5.20) and (5.21) hold. Since $|x_n - y_n| \geq \delta$, either $x_n - y_n \geq \delta$ or $x_n - y_n \leq -\delta$.

Let $x_n - y_n \geq \delta$. Then for $m \geq l'$, we have

$$x_m > x_n - \frac{\delta}{3}$$
$$\geq y_n + \frac{2\delta}{3}$$
$$= y_n + \frac{\delta}{3} + \frac{\delta}{3}$$
$$> y_m + \frac{\delta}{3}.$$

This implies that $\hat{y} < \hat{x}$. Now if $x_n - y_n \leq -\delta$, then $y_n - x_n \geq \delta$. Then by a similar argument as above we get $\hat{x} < \hat{y}$.

We define $\hat{x} \leq \hat{y}$ if $\hat{x} < \hat{y}$ or $\hat{x} = \hat{y}$. We can check that $\leq$ is a partial order relation on $\hat{\mathbb{Q}}$ and $(\hat{\mathbb{Q}}, +, \cdot)$ is an ordered field.

Theorem 5.4.2 (Archimedean property). *Let $\hat{x}, \hat{y} > \hat{0}$. Then there exists $\hat{m} \in \hat{\mathbb{Q}}$ such that $\hat{m}\hat{x} > \hat{y}$, where $\hat{m}$ is the equivalence class of the constant sequence $m \in \mathbb{N}$.*

Proof. We have to show that there are a positive rational number δ and $k, m \in \mathbb{N}$ such that

$$n \geq k \Rightarrow mx_n > y_n + \delta.$$

We will prove this by contradiction. On the contrary, suppose that for every rational $\delta > 0$ and for all $m, k \in \mathbb{N}$, there is $n \geq k$ such that

$$mx_n \leq y_n + \delta. \tag{5.22}$$

Since a Cauchy sequence is bounded, there is $M > 0$ such that $y_n < M$ for all $n \in \mathbb{N}$. This implies that

$$mx_n \leq y_n + \delta < M + \delta.$$

Let $\epsilon > 0$. For a fixed $\delta = \delta_1 > 0$, choose $m_1 \in \mathbb{N}$ such that

$$\frac{M + \delta_1}{m_1} < \frac{\epsilon}{2}.$$

Therefore, for these $\delta_1 > 0$ and $m_1 \in \mathbb{N}$ and for all $k \in \mathbb{N}$, there is $n \geq k$ such that

$$x_n < \frac{M + \delta_1}{m_1} < \frac{\epsilon}{2}.$$

Since (x_n) is a Cauchy sequence, there is $k_1 \in \mathbb{N}$ such that

$$r, s \geq k_1 \Rightarrow |x_r - x_s| < \frac{\epsilon}{2}. \tag{5.23}$$

By the contrary assumption there is $l \geq k_1$ such that

$$m_1 x_l \leq y_l + \delta_1.$$

Therefore

$$x_l \leq \frac{y_l + \delta_1}{m_1} < \frac{M + \delta_1}{m_1} < \frac{\epsilon}{2}.$$

By equation (5.23), for $r \geq k_1$, we have

$$x_r < x_l + \frac{\epsilon}{2} < \frac{\epsilon}{2} + \frac{\epsilon}{2} = \epsilon.$$

This shows that $x_r \to 0$. This is a contradiction as $\hat{x} > \hat{0}$. ☐

Now we prove that $\hat{\mathbb{Q}}$ satisfies the least upper bound property. Before that, we prove the following:

Proposition 5.4.3. *Let $\mathbb{F}$ be an ordered field such that Archimedean property holds in $\mathbb{F}$. Let (x_n) be an increasing and bounded above sequence in $\mathbb{F}$. Then (x_n) is a Cauchy sequence.*

Proof. Let $l \in \mathbb{F}$ be such that $x_n \leq l$ for all $n \in \mathbb{N}$. Let $\epsilon > 0$. Then either there is $k \in \mathbb{N}$ such that $l - \epsilon < x_k \leq l$, or $x_n \leq l - \epsilon$ for all $n \in \mathbb{N}$.

First, suppose that there is $k \in \mathbb{N}$ such that $l - \epsilon < x_k \leq l$. Since (x_n) is an increasing sequence, for all $n \geq k$, we have

$$l - \epsilon < x_k \leq x_n \leq l.$$

This implies that

$$r, s \geq k \Rightarrow |x_r - x_s| < \epsilon.$$

Now suppose that $x_n \leq l - \epsilon$ for all $n \in \mathbb{N}$. Then we replace l by $l - \epsilon$. By the Archimedean property we can find $m \in \mathbb{N}$ such that

$$l - m\epsilon < x_k \leq l - (m - 1)\epsilon$$

for some $k \in \mathbb{N}$ and $x_n \leq l - (m-1)\epsilon$ for all $n \in \mathbb{N}$. Let $L = l - (m-1)\epsilon$. Then $L - \epsilon < x_k \leq L$. By the first case we get

$$r, s \geq k \Rightarrow |x_r - x_s| < \epsilon.$$

This shows that (x_n) is a Cauchy sequence. ☐

Corollary 5.4.4. *Let* $\mathbb{F}$ *be an ordered field such that the Archimedean property holds in* $\mathbb{F}$. *Let* (x_n) *be a decreasing and bounded below sequence in* $\mathbb{F}$. *Then* (x_n) *is a Cauchy sequence.*

Theorem 5.4.5. *The least upper bound property holds in* $\hat{\mathbb{Q}}$.

Proof. Let S be a nonempty subset in $\hat{\mathbb{Q}}$ that is bounded above by l (to avoid the complexity of the symbol, we drop the hat of the elements). Let $a \in S$. Now we will inductively define the sequences in $\hat{\mathbb{Q}}$. Let $x_1 = l$ and $y_1 = a$. Suppose that we have defined the terms x_n and y_n. Then we define $x_{n+1} = \frac{x_n + y_n}{2}$ and $y_{n+1} = y_n$ if $\frac{x_n + y_n}{2}$ is an upper bound of S; otherwise, we define $x_{n+1} = x_n$ and $y_{n+1} = \frac{x_n + y_n}{2}$ if $\frac{x_n + y_n}{2}$ is not an upper bound of S. Since (y_n) is increasing and bounded above by l, by Proposition 5.4.3 (y_n) is a Cauchy sequence. Also, since (x_n) is a decreasing sequence bounded below by a, by Corollary 5.4.4 (x_n) is a Cauchy sequence. Since $\hat{\mathbb{Q}}$ is complete, $x_n \to x$ in $\hat{\mathbb{Q}}$.

Next, we claim that $y_n \to x$. First, suppose that $\frac{x_n + y_n}{2}$ is an upper bound of S. Then

$$x_{n+1} - y_{n+1} = \frac{x_n + y_n}{2} - y_n = \frac{x_n - y_n}{2}. \tag{5.24}$$

Now suppose that $\frac{x_n + y_n}{2}$ is not an upper bound of S. Then

$$x_{n+1} - y_{n+1} = x_n - \frac{x_n + y_n}{2} = \frac{x_n - y_n}{2}. \tag{5.25}$$

By equations (5.24) and (5.25) we get

$$x_n - y_n = \frac{1}{2^{n-1}}(l - a).$$

This implies that $\lim_{n \to \infty}(x_n - y_n) = 0$. Therefore

$$\lim_{n \to \infty} y_n = \lim_{n \to \infty} x_n = x.$$

Next, we claim that x is an upper bound of S. On the contrary, suppose that there exists $a \in S$ such that $x < a$. Then $a - x > 0$. Note that $x_n - x$ is decreasing and $x_n - x \to 0$. Then there exists $k \in \mathbb{N}$ such that $x_k - x < a - x$. This implies that $x_k < a$. This is a contradiction as x_k is an upper bound of S.

Finally, we claim that x is the least upper bound of S. Let $\epsilon > 0$. Then $x - \epsilon < x$. Note that (y_n) is increasing and $y_n \to x$. Then there exists $k \in \mathbb{N}$ such that

$$x - \epsilon < y_k \leq x.$$

Since (y_n) is increasing,

$$n \geq k \Rightarrow x - \epsilon < y_n.$$

Also, note that y_n is not an upper bound of S for each $n \in \mathbb{N}$. There exists $z_n \in S$ such that $y_n \leq z_n$. Therefore for all $n \geq k$, we have $x - \epsilon < z_n$. Since x is an upper bound of S,

$$x - \epsilon < z_n \le x.$$

This shows that x is the least upper bound of S. $\qquad\qquad\qquad\qquad$ □

Thus we have proved that $\hat{Q}$ is a complete ordered field. Since a complete ordered field is isomorphic to $\mathbb{R}$, the completion of $\mathbb{Q}$ with the usual metric is $\mathbb{R}$.

Let p be a prime. Consider the metric d_p on $\mathbb{Q}$ as given in Example 2.1.12. The completion of $\mathbb{Q}$ with this metric is called the *p-adic completion* of $\mathbb{Q}$. For more detail, see [5].

5.5 Baire category theorem

In this section, we will prove the Baire category theorem and obtain some of its applications. This concept of category was introduced by R. Baire in his thesis. He divided the subsets of a metric space into two types or categories.

Definition 5.5.1. Let A and B be two sets in a metric space X. Then A is said to be dense in B if $A \subseteq B$ and every open set U that intersects B intersects A.

Proposition 5.5.2. *Let A and B be two sets in a metric space X. Then A is dense in B if and only if $B \subseteq \mathrm{Cl}\,A$.*

Proof. Suppose that A is dense in B. Let $x \in B$. If $x \in A$, then $x \in \mathrm{Cl}\,A$. Now suppose that $x \notin A$. Then every open set around x intersects B and A. This implies that x is a limit point of A. Therefore $x \in \mathrm{Cl}\,A$. This shows that $B \subseteq \mathrm{Cl}\,A$.

Conversely, suppose that $B \subseteq \mathrm{Cl}\,A$. Let U be an open set such that $U \cap B \ne \emptyset$. This implies that $U \cap \mathrm{Cl}\,A \ne \emptyset$. Suppose that $x \in U \cap \mathrm{Cl}\,A$. If $x \in A$, then $U \cap A \ne \emptyset$. Suppose that $x \notin A$. Since $x \in U$, there is an open set V such that $x \in V \subseteq U$. Note that x is a limit point of A. This implies that $V \cap A \ne \emptyset$. Therefore $U \cap A \ne \emptyset$. Hence A is dense in B. $\quad$ □

Definition 5.5.3. A set A in a metric space X is called a *nowhere dense set* if every nonempty open set U in X contains a nonempty open set V such that $V \cap A = \emptyset$.

Example 5.5.4. Every finite set and a line in $\mathbb{R}^n$ are nowhere dense sets.

Example 5.5.5. The set $\{\frac{1}{n} \mid n \in \mathbb{N}\}$ is nowhere dense set in $\mathbb{R}$.

Example 5.5.6. No nonempty set in the discrete metric space is a nowhere dense set.

We can check that a subset of a nowhere dense set is a nowhere dense set. Also, the closure of a nowhere dense set is a nowhere dense set. If $f : X \to Y$ is a homeomorphism and A is a nowhere dense set, then $f(A)$ is a nowhere dense set.

Proposition 5.5.7. *A finite union of nowhere dense sets is a nowhere dense set.*

Proof. Let $A_1, \ldots, A_n$ be nowhere dense sets. We can suppose that each A_i is nonempty. Let U be a nonempty open set in X. Since A_1 is a nowhere dense set, there is a nonempty open set $V_1 \subseteq U_1$ such that $V_1 \cap A_1 = \emptyset$.

Since A_2 is a nowhere dense set, there is a nonempty open set $V_2 \subseteq V_1$ such that $V_2 \cap A_2 = \emptyset$. Inductively, we get a nonempty open set $V_n \subseteq V_{n-1}$ such that $V_n \cap A_n = \emptyset$. Since

$$V_n \subseteq V_{n-1} \subseteq \cdots \subseteq V_1 \subseteq U$$

and $V_i \cap A_i \neq \emptyset$, $V_n \cap (\bigcup A_i) = \emptyset$. Therefore $\bigcup A_i$ is a nowhere dense set. □

Example 5.5.8. A countable union of nowhere dense sets need not be a nowhere dense set. For each $n \in \mathbb{N}$, define the set

$$A_n = \left\{ \frac{m}{n} \mid m \in \mathbb{Z} \right\}.$$

Then each A_n is a nowhere dense set. Note that $\bigcup_{n \in \mathbb{N}} A_n = \mathbb{Q}$, which is not a nowhere dense set.

Theorem 5.5.9. *Let A be a set in a metric space X. Then the following statements are equivalent:*
(i)   *A is a nowhere dense set;*
(ii) *$\text{Int Cl} A = \emptyset$;*
(iii) *$\text{Cl} A$ contains no nonempty open set in X;*
(iv) *Every nonempty open set U in X contains a nonempty open set V such that $V \cap \text{Cl} A = \emptyset$.*

Proof. (ii) $\Rightarrow$ (iii) and (iv) $\Rightarrow$ (i) are obvious. We will prove (i) $\Rightarrow$ (ii) and (iii) $\Rightarrow$ (iv).

Suppose (i) holds. On the contrary, suppose that $\text{Int Cl} A \neq \emptyset$. Let $U = \text{Int Cl} A$. By the assumption there is a nonempty open set $V \subseteq U$ such that $V \cap A = \emptyset$. Let $x \in V$. Then $x \notin A$. Since $V \cap A = \emptyset$, x cannot be a limit point of A. This implies that $x \notin A \cup D(A) = \text{Cl} A$. This is a contradiction as

$$x \in V \subseteq U = \text{Int Cl} A \subseteq \text{Cl} A.$$

This proves (ii). Now assume (iii). Let U be a nonempty open set in X. Then $U \cap (X \setminus \text{Cl} A) \neq \emptyset$, since otherwise, $U \subseteq \text{Cl} A$. Since $U \cap (X \setminus \text{Cl} A)$ is open, for each $x \in U \cap (X \setminus \text{Cl} A)$, there is an open ball $B(x, r)$ contained in $U \cap (X \setminus \text{Cl} A)$. Then $B(x, r) \cap \text{Cl} A = \emptyset$. This proves (iv). □

Proposition 5.5.10. *A set A in a metric space X is a nowhere dense set if and only if $X \setminus \text{Cl} A$ is dense in X.*

Proof. Suppose that A is a nowhere dense set. Let $x \in X$. Suppose that $x \in \text{Cl} A$. Since $\text{Int Cl} A = \emptyset$, x in not an interior point of $\text{Cl} A$. This implies that for each $r > 0$, $B(x, r) \nsubseteq \text{Cl} A$. This equivalently means that for each $r > 0$,

$$B(x, r) \cap (X \setminus \text{Cl} A) \neq \emptyset.$$

This implies that x is a limit point of $X \setminus \text{Cl} A$. This shows that $X \setminus \text{Cl} A$ is dense in X.

Conversely, suppose that $X \setminus \mathrm{Cl}\,A$ is dense in X. Let $x \in \mathrm{Cl}\,A$. Then for each $r > 0$,

$$B(x,r) \cap (X \setminus \mathrm{Cl}\,A) \neq \emptyset.$$

Therefore $B(x,r) \not\subseteq \mathrm{Cl}\,A$. Hence x is not an interior point of $\mathrm{Cl}\,A$. Since $x \in \mathrm{Cl}\,A$ is arbitrary,

$$\mathrm{Int}\,\mathrm{Cl}\,A = \emptyset.$$

Therefore A is a nowhere dense set. ☐

Definition 5.5.11. A set A in a metric space X is called a set of first category or meager if it can be represented as a countable union of nowhere dense sets.

Definition 5.5.12. A set A in a metric space X is called a set of second category if it is not of first category.

Example 5.5.13. Every nowhere dense set is of first category.

Example 5.5.14. The set of rational numbers $\mathbb{Q}$ is of first category in $\mathbb{R}$.

Consider an open ball $B(x,r)$ in a metric space X. Let $y \in \mathrm{Cl}\,B(x,r)$. Then we claim that $d(x,y) \leq r$, that is, $y \in D(x,r)$.

Suppose that $y \notin D(x,r)$. Then $d(x,y) > r$. Let $z \in B(y, d(x,y) - r) \cap B(x,r)$. Then

$$d(z,y) < d(x,y) - r \quad \text{and} \quad d(z,x) < r.$$

Now

$$d(x,z) + d(z,y) \geq d(x,y) > d(z,y) + r.$$

This implies that $d(x,z) > r$, a contradiction. Therefore

$$B(y, d(x,y) - r) \cap B(x,r) = \emptyset.$$

Hence $y \notin \mathrm{Cl}\,B(x,r)$. This shows that

$$y \in \mathrm{Cl}\,B(x,r) \Rightarrow y \in D(x,r).$$

Thus $\mathrm{Cl}\,B(x,r) \subseteq D(x,r)$.

Let U be a nonempty open set, and let $x \in U$. Then there exists an open ball $B(x,r) \subseteq U$. Now

$$B\left(x, \frac{r}{2}\right) \subseteq D\left(x, \frac{r}{2}\right) \subseteq B(x,r) \subseteq U.$$

Therefore we can find an open ball in U whose closure lies inside U.

Theorem 5.5.15 (Baire category theorem). *A nonempty open set in a complete metric space is of second category.*

Proof. Let X be a complete metric space, and let U be a nonempty open set in X. Let $\{A_n \mid n \in \mathbb{N}\}$ be any countable collection of nowhere dense sets. Then we can find an open ball $B(x_1, r_1)$ with $0 < r_1 < 1$ such that

$$\mathrm{Cl}\, B(x_1, r_1) \subseteq U \quad \text{and} \quad \mathrm{Cl}\, B(x_1, r_1) \cap A_1 = \emptyset.$$

Suppose that we have found $B(x_n, r_n)$ with $0 < r_n < \frac{1}{n}$. Then we can find an open ball $B(x_{n+1}, r_{n+1})$ with $0 < r_{n+1} < \frac{1}{n+1}$ such that

$$\mathrm{Cl}\, B(x_{n+1}, r_{n+1}) \subseteq B(x_n, r_n) \quad \text{and} \quad \mathrm{Cl}\, B(x_{n+1}, r_{n+1}) \cap A_{n+1} = \emptyset.$$

In turn, we get a sequence (x_n) in X. We can check that (x_n) is a Cauchy sequence. Since X is complete, $x_n \to x$ in X. Note that the subsequence $(x_{n+1}, x_{n+2}, \dots)$ of (x_n) lies in $B(x_{n+1}, r_{n+1})$. Since a subsequence of a convergent sequence converges to the same point,

$$x \in \mathrm{Cl}\, B(x_{n+1}, r_{n+1}) \subseteq B(x_n, r_n).$$

Therefore $x \in B(x_n, r_n)$ for all $n \in \mathbb{N}$. Hence $x \notin \bigcup_{n=1}^{\infty} A_n$. Since $x \in U$, $U \neq \bigcup_{n=1}^{\infty} A_n$. This shows that U is of second category. □

Corollary 5.5.16. *Every complete metric space is of second category.*

Corollary 5.5.17. *Let X be a complete metric space. Then every countable union of closed nowhere dense sets in X has an empty interior.*

Proof. Let $\{A_n \mid n \in \mathbb{N}\}$ be a countable family of closed nowhere dense sets. On the contrary, suppose that

$$\mathrm{Int}\left(\bigcup_{n=1}^{\infty} A_n \right) \neq \emptyset.$$

Let $U = \mathrm{Int}(\bigcup_{n=1}^{\infty} A_n)$. Consider the collection $\{A_n \cap U \mid n \in \mathbb{N}\}$. Since a subset of nowhere dense sets is a nowhere dense set, $A_n \cap U$ is a nowhere dense set. Then

$$U = U \cap \bigcup_{n=1}^{\infty} A_n = \bigcup_{n=1}^{\infty} (U \cap A_n).$$

This implies that U is of first category, a contradiction. □

Corollary 5.5.18. *Let X be a complete metric space. Then a countable intersection of open dense sets is dense.*

Proof. Let $\{U_n \mid n \in \mathbb{N}\}$ be a countable collection of open dense sets. Then $X \setminus U_n$ is a closed and nowhere dense set. This implies that

$$\text{Int}\left(\bigcup_{n=1}^{\infty} X \setminus U_n\right) = \emptyset.$$

This shows that $\text{Cl} \bigcap_{n=1}^{\infty} U_n = X$. $\qquad\square$

Corollary 5.5.19. *Let X be a complete metric space. Let $\{A_n \mid n \in \mathbb{N}\}$ be a countable collection of closed sets with empty interior. Then*

$$\text{Int} \bigcup_{n=1}^{\infty} A_n = \emptyset.$$

Proof. If A_n has an empty interior, then $X \setminus A_n$ is dense. Also, note that $X \setminus A_n$ is open. Then $\bigcap_{n=1}^{\infty} X \setminus A_n$ is dense. Hence $\bigcup_{n=1}^{\infty} A_n$ has an empty interior. $\qquad\square$

Corollary 5.5.20. *Let X be a complete metric space. Then every set of first category has an empty interior.*

Proof. Let A be a set of first category in X. Note that a subset of a set of first category is also of first category. If A has a nonempty interior, then the interior as a subset of A must be of first category, a contradiction. $\qquad\square$

Corollary 5.5.21. *Let X be a complete metric space, and let A be of first category. Then $X \setminus A$ is dense.*

We can construct a function $f : \mathbb{R} \to \mathbb{R}$ that is discontinuous only at finitely many points. For example, we can consider a continuous map $g : \mathbb{R} \to \mathbb{R}$ and define the function $f : \mathbb{R} \to \mathbb{R}$ by

$$f(x) = \begin{cases} g(x) & \text{if } x \neq a, \\ b & \text{if } x = a, \end{cases}$$

where $b \neq f(a)$. Then we can check that f is discontinuous at only one point. Consider the function $f : \mathbb{R} \to \mathbb{R}$ defined by

$$f(x) = \begin{cases} \frac{1}{n} & \text{if } x = \frac{m}{n} \text{ with } \gcd(m, n) = 1, \\ 0 & \text{if } x \text{ is irrational}, \end{cases}$$

where $\gcd(m, n)$ denotes the greatest common divisor of m and n. We can check that f is continuous at irrationals and discontinuous at rationals. We will see that there does not exist a function $f : \mathbb{R} \to \mathbb{R}$ that is continuous at rationals and discontinuous at irrationals.

Definition 5.5.22. A set A in a metric space is called a G_δ-set if it is a countable intersection of open sets.

Example 5.5.23. Every open set is a G_δ-set. Since $(a, b] = \bigcap_{n=1}^{\infty}(a, b + \frac{1}{n})$, $(a, b]$ is a G_δ-set.

Proposition 5.5.24. *Every closed set in a metric space is a G_δ-set.*

Proof. Let X be a metric space, and let F be a closed set in X. For each $n \in \mathbb{N}$, define

$$F_n = \bigcup_{x \in F} B\left(x, \frac{1}{n}\right).$$

Note that F_n is an open set and $F \subseteq F_n$ for all $n \in \mathbb{N}$. This implies that $F \subseteq \bigcap F_n$.

Let $y \in \bigcap F_n$. Let $\epsilon > 0$. By the Archimedean property there is $m \in \mathbb{N}$ such that $\frac{1}{m} < \epsilon$. Since $y \in F_m$, there is $x \in F$ such that

$$d(x, y) < \frac{1}{m} < \epsilon. \tag{5.26}$$

We claim that $y \in F$. If $y \notin F$, then $x \neq y$. Then by equation (5.26) we see that y is a limit point of F. Since F is closed, $y \in F$, a contradiction. Since $y \in \bigcap F_n$ is arbitrary, $\bigcap F_n \subseteq F$. This shows that $F = \bigcap F_n$. Hence F is a G_δ-set. $\square$

Proposition 5.5.25. *Let X be a complete metric space. Let A be a subset of X that is countable and dense and has no isolated point. Then A is not a G_δ-set.*

Proof. Let $a \in A$. Then $X \setminus \{a\}$ is open in X. Since A has no isolated point, $\{a\}$ is not open. This implies that $X \setminus \{a\}$ is not closed. Therefore $\text{Cl}\, X \setminus \{a\}$ properly contains $X \setminus \{a\}$. Therefore $X \setminus \{a\}$ is dense in X. Note that

$$X \setminus A = \bigcap_{a \in A}(X \setminus \{a\}).$$

Since A is countable, $X \setminus A$ is a G_δ-set. Also, since $X \setminus \{a\}$ is dense, by the Baire category theorem $X \setminus A$ is dense.

On the contrary, suppose that A is a G_δ-set. Then there is a countable collection

$$\{U_n \mid n \in \mathbb{N}\}$$

of open sets such that $A = \bigcap U_n$. Since $A \subseteq U_n$ for all $n \in \mathbb{N}$ and A is dense, U_n is dense. Note that

$$\{U_n \mid n \in \mathbb{N}\} \cup \{X \setminus \{a\} \mid a \in A\}$$

is a countable collection of open dense sets. By the Baire category theorem the intersection of these sets is a dense set. This is a contradiction as

$$\left(\bigcap U_n\right) \cap \bigcap_{a \in A}(X \setminus \{a\}) = A \cap (X \setminus A) = \emptyset. \qquad \square$$

Let U be an open set in a metric space X. For a real-valued function $f : U \to \mathbb{R}$, we defines the oscillation of f on U as

$$\mathrm{osc}_f(U) = \sup\{f(x) \mid x \in U\} - \inf\{f(x) \mid x \in U\}.$$

We also define the oscillation of f at $a \in \mathrm{Cl}\, U$ as

$$\mathrm{osc}_f(a) = \inf\{\mathrm{osc}_f(B(a,r) \cap U) \mid r > 0\}.$$

Let X and Y be metric spaces. We can generalize the notion of oscillation for a function $f : A \subseteq X \to Y$ at $a \in \mathrm{Cl}\, A$ as follows:

$$\mathrm{osc}_f(a) = \inf\{\mathrm{diam}(f(A \cap B(a,r))) \mid r > 0\}.$$

Proposition 5.5.26. *Let X and Y be metric spaces, and let $A \subseteq X$. Let $f : A \to Y$ be a map. Then f is continuous at $a \in A$ if and only if $\mathrm{osc}_f(a) = 0$.*

Proof. The oscillation $\mathrm{osc}_f(a) = 0$ if and only if for each $\epsilon > 0$, there is an open ball $B(a,r)$ such that $\mathrm{diam}(f(A \cap B(a,r))) < \epsilon$. This equivalently means that

$$d(x,a) < r \Rightarrow d(f(x),f(a)) < \epsilon.$$

Hence $\mathrm{osc}_f(a) = 0$ if and only if f is continuous at $a \in A$. $\qquad \square$

Proposition 5.5.27. *Let X and Y be metric spaces. Let $A \subseteq X$, and let $f : A \to Y$. Then the set*

$$Z = \{x \in \mathrm{Cl}\, A \mid \mathrm{osc}_f(x) = 0\}$$

is a G_δ-set in $\mathrm{Cl}\, A$.

Proof. For each $n \in \mathbb{N}$, define

$$Z(n) = \left\{x \in \mathrm{Cl}\, A \;\middle|\; \mathrm{osc}_f(x) < \frac{1}{n}\right\}.$$

Note that $Z = \bigcap_{n=1}^{\infty} Z(n)$. We claim that each $Z(n)$ is an open set in $\mathrm{Cl}\, A$. Note that

$$\mathrm{osc}_f(x) = \inf\{\mathrm{diam}(f(A \cap B(x,r))) \mid r > 0\} < \frac{1}{n}$$

if and only if there is $s > 0$ such that

$$\mathrm{diam}(f(A \cap B(x,s))) < \frac{1}{n}.$$

Therefore

$$Z(n) = \left\{ x \in \mathrm{Cl}\, A \; \middle| \; \mathrm{osc}_f(x) < \frac{1}{n} \right\}$$

$$= \left\{ x \in \mathrm{Cl}\, A \; \middle| \; \text{there is } s > 0 \text{ such that } \mathrm{diam}(f(A \cap B(x, s))) < \frac{1}{n} \right\}$$

$$= \mathrm{Cl}\, A \cap \left\{ x \in X \; \middle| \; \text{there is } s > 0 \text{ such that } \mathrm{diam}(f(A \cap B(x, s))) < \frac{1}{n} \right\}$$

$$= \mathrm{Cl}\, A \cap \bigcup_{x \in X} \left\{ B(x, r) \; \middle| \; \mathrm{diam}(f(A \cap B(x, r))) < \frac{1}{n} \right\}.$$

This implies that $Z(n)$ is open in $\mathrm{Cl}\, A$. □

Corollary 5.5.28. *Let X and Y be metric spaces, and let $f : X \to Y$. Then*

$$Z = \{ x \in \mathrm{Cl}\, A \mid f \text{ is continuous at } x \}$$

is a G_δ-set.

Proof. By Proposition 5.5.26

$$Z = \{ x \in X \mid \mathrm{osc}_f(x) = 0 \}.$$

By Proposition 5.5.27, Z is a G_δ-set. □

Corollary 5.5.29. *Let X and Y be metric spaces with Y a complete metric space. Let $A \subseteq X$, and let $f : X \to Y$. Then*

$$Z = \{ x \in \mathrm{Cl}\, A \mid \mathrm{osc}_f(x) = 0 \}$$

is a G_δ-set in X.

Proof. By Proposition 5.5.27, Z is a G_δ-set in $\mathrm{Cl}\, A$. By Proposition 5.5.24, $\mathrm{Cl}\, A$ is a G_δ-set in X. Therefore Z is a G_δ-set in X. □

By Proposition 5.5.25, $\mathbb{Q}$ is not a G_δ-set. Therefore there is no map $f : \mathbb{R} \to \mathbb{R}$ such that f is continuous at the points of $\mathbb{Q}$ and discontinuous at the points of $\mathbb{R} \setminus \mathbb{Q}$. Now it is natural to ask if there is a function that is continuous exactly on a G_δ-set. We have the following:

Theorem 5.5.30 (Sung Soo Kim). *Let X be a metric space without isolated points, and let A be a G_δ-set in X. Then there is a function that is continuous exactly on A.*

Proof. Since A is a G_δ-set, $X \setminus A$ is a countable union of closed sets A_n, $n \in \mathbb{N}$. Let $F_1 = A_1$. Suppose we have obtained F_n. Then we define $F_{n+1} = F_n \cup A_{n+1}$. Note that each F_n is closed and $F_n \subseteq F_{n+1}$ for all $n \in \mathbb{N}$. Also, note that $X \setminus A$ is a countable union of closed sets F_n. Define the function $g : X \to \mathbb{R}$ by $g(x) = \sum_{n \in K} \frac{1}{2^n}$, where $K = \{ n \in \mathbb{N} \mid x \in F_n \}$.

Note that if $x \in A$, then $n \to \infty$, that is the right-hand side of $g(x)$ converges to 0. By Proposition 3.2.30 we get a dense set B in X whose complement is also dense in X. Let

$$f(x) = g(x)\left(\chi_B(x) - \frac{1}{2}\right) = \begin{cases} \frac{g(x)}{2} & \text{if } x \in B, \\ \frac{-g(x)}{2} & \text{if } x \in X \setminus B, \end{cases}$$

where $\chi_B(x)$ is the indicator function on B. We will first observe that f is discontinuous at every point of $X \setminus A$. Let $x \in X \setminus A$. Then either $x \in \text{Int}(X \setminus A)$, or $x \in (X \setminus A) \cap D(A)$.

Let $x \in \text{Int}(X \setminus A)$. Then every open ball around x contains a point of B and a point of $X \setminus B$. This implies that every open ball around x contains a point at which the sign of f is different from the sign of $f(x)$. Therefore f is not continuous at the interior point of $X \setminus A$.

Let $x \in (X \setminus A) \cap D(A)$. Then every open ball around x contains a point y of A. Note that $f(x) \neq 0$ and $f(y) = 0$. Therefore f is not continuous at any point of $(X \setminus A) \cap D(A)$. Hence f is not continuous at any point of $X \setminus A$. By the sequential criterion of continuity we can observe that the function f is continuous on A. □

Let us revisit Remark 5.2.14. Suppose we have a subspace Y of a complete metric space X that is not complete. It is a natural question whether there is a complete metric on Y that is topologically equivalent to the subspace metric on Y. We have the following:

Theorem 5.5.31. *Let (X, d) be a complete metric, and let Y be a G_δ-set in X. Then there is a complete metric ρ on Y that is topologically equivalent to the subspace metric d on Y.*

Proof. Since Y is a G_δ-set, $Y = \bigcap_{i=1}^{\infty} U_i$, where $\{U_i \mid i \in \mathbb{N}\}$ is a countable family of open sets. Let $y_1, y_2 \in Y$. Define $\rho : Y \times Y \to \mathbb{R}$ by

$$\rho(y_1, y_2) = d(y_1, y_2) + \sum_{i=1}^{\infty} \frac{1}{2^i} \min\left\{1, \left|\frac{1}{d(y_1, X \setminus U_i)} - \frac{1}{d(y_2, X \setminus U_i)}\right|\right\}.$$

We can check that ρ is a complete metric on Y that is topologically equivalent to the metric d on Y. □

In Theorem 5.5.31, we considered the subspace Y to be a G_δ-set. This is a necessary requirement as observed below. We leave its proof as an exercise, which can be obtained by a little effort.

Theorem 5.5.32. *Let (X, d) be a metric, and let Y be a subspace of X. Suppose there is a complete metric ρ on Y that is topologically equivalent to the subspace metric d on Y. Then Y is a G_δ-set in X.*

Let us see another application of the Baire category theorem.

Let X and Y be metric spaces. Let $a \in X$. Then a collection $\mathcal{F}$ of maps from X to Y is called pointwise bounded if the set $\{f(a) \mid f \in \mathcal{F}\}$ is bounded in Y.

A collection $\mathcal{F}$ of maps from X to Y is called uniformly bounded on $A \subseteq X$ if the set $\{f(x) \mid f \in \mathcal{F}, x \in A\}$ is bounded in Y.

Theorem 5.5.33 (Osgood theorem). *Let X and Y be metric spaces with X a complete metric space. Let a collection $\mathcal{F}$ of continuous maps from X to Y be pointwise bounded for each $x \in X$. Then there is a nonempty open set U such that $\mathcal{F}$ is uniformly bounded on U.*

Proof. Let $a \in X$. For each $n \in \mathbb{N}$, define

$$A_n = \{x \in X \mid f(x) \in \mathrm{Cl}\, B(a, n) \text{ for all } f \in \mathcal{F}\}.$$

Then

$$A_n = \bigcap_{f \in \mathcal{F}} f^{-1}(\mathrm{Cl}\, B(a, n)).$$

Since f is continuous, A_n is closed. Since $\mathcal{F}$ is pointwise bonded, for each $x \in X$, there is $n \in \mathbb{N}$ such that

$$\{f(x) \mid f \in \mathcal{F}\} \subseteq B(a, n) \subseteq \mathrm{Cl}\, B(a, n).$$

This shows that $X = \bigcup_{n=1}^{\infty} A_n$. By Corollary 5.5.19 there is $m \in \mathbb{N}$ such $\mathrm{Int}\, A_m \neq \emptyset$. Let $U = \mathrm{Int}\, A_m$. By the definition of A_m,

$$\{f(x) \mid f \in \mathcal{F}, \, x \in U\} \subseteq \mathrm{Cl}\, B(a, m).$$

Hence $\mathcal{F}$ is uniformly bounded on U. □

We have the experience in calculus that a continuous function whose graph has a corner at some point is not differentiable at that point. It is natural to ask whether there are continuous functions that are nowhere differentiable. At first it seems impossible, but the Baire category theorem ensures that the collection of such functions is dense in the space of continuous functions with the supremum metric d_∞.

By a continuous piecewise linear function $f : [0, 1] \to \mathbb{R}$ we mean a continuous function whose graph contains finitely many line segments. It is left as an exercise to check that the set of all continuous piecewise linear functions is dense in $(C[0, 1], d_\infty)$.

The function $\psi : \mathbb{R} \to \mathbb{R}$ defined by

$$\psi(x) = \min\{x - [x], [x] + 1 - x\}$$

is called the triangular sawtooth function, where $[x]$ denotes the greatest integer function. We will use the result proved later that every sequence in the closed interval $[a, b]$ has a convergent subsequence.

Theorem 5.5.34. *The set of all nowhere differentiable functions is dense in $(C[0, 1], d_\infty)$.*

Proof. Let us consider the complement of the set of all nowhere differentiable functions in $C[0,1]$. If a function $f \in C[0,1]$ is differentiable at some point $x \in [0,1)$, then the right-hand difference quotients

$$\frac{f(x+h)-f(x)}{h} \quad (0 < h < 1-x)$$

are bounded. For each $n \in \mathbb{N}$, define the set

$$A_n = \left\{ f \in C[0,1] \,\middle|\, \text{for some } x \in \left[0, 1-\frac{1}{n}\right], \right.$$

$$\left. |f(x+h)-f(x)| \leq nh \text{ for all } 0 < h < 1-x \right\}.$$

Let $A = \bigcup A_n$. Then A is the set of all continuous functions $f \in C[0,1]$ that have the right-hand derivative at some point in $[0,1)$. The complement of A in $C[0,1]$ is the set of all nowhere differentiable functions in $C[0,1]$.

We first show that each A_n is closed. Let $f \in \text{Cl}\, A_n$. Then there is a sequence (f_m) in A_n such that $f_m \to f$. Since $f_m \in A_n$, there is a point $x_m \in [0, 1-\frac{1}{n}]$ such that

$$|f_m(x_m + h) - f(x_m)| \leq nh \quad \text{for all } 0 < h < 1 - x_m.$$

This gives a sequence (x_m) in $[0, 1-\frac{1}{n}]$. Then it has a convergent subsequence that converges to some point $a \in [0, 1-\frac{1}{n}]$. We may reindex to assume that $x_m \to a$.

Let $0 < h < 1 - a$. Then we claim that there is $k \in \mathbb{N}$ such that $0 < h < 1 - x_k$. On the contrary, suppose that for each $n \in \mathbb{N}$, there is $M > n$ such that $1 - x_N \leq h$. This would imply that $1 - a \leq h$, a contradiction. Now by the triangle inequality we have

$$\begin{aligned}
|f(a+h)-f(a)| &\leq |f(a+h)-f(x_k + h)| + |f(x_k + h) - f_k(x_k + h)| \\
&\quad + |f_k(x_k + h) - f_k(x_k)| + |f_k(x_k) - f(x_k)| \\
&\quad + |f(x_k) - f(a)| \\
&\leq |f(a+h)-f(x_k + h)| + d_\infty(f, f_k) + nh \\
&\quad + d_\infty(f_k, f) + |f(x_k) - f(a)|.
\end{aligned}$$

Since f is continuous, taking the limit as $k \to \infty$, we have

$$|f(a+h) - f(a)| \leq nh.$$

This implies that $f \in A_n$. Therefore A_n is closed.

Now we will prove that each $C[0,1] \setminus A_n$ is a dense set. Note that the set of all continuous piecewise linear functions is dense in $C[0,1]$. Let $\epsilon > 0$. Then for any $f \in C[0,1]$, there is a continuous piecewise linear function $h \in B(f, \frac{\epsilon}{2})$. We will show that the open ball $B(h, \frac{\epsilon}{2})$ intersects $C[0,1] \setminus A_n$ nontrivially. This will jointly imply that $B(f, \epsilon)$ intersects $C[0,1] \setminus A_n$ nontrivially.

Let L be the maximum slope of the line segment of the piecewise linear function h. Let l be a real number such that

$$l\frac{\epsilon}{2} > L + n. \tag{5.27}$$

Let $g(x) = h(x) + \frac{\epsilon}{2}\psi(lx)$, where ψ is the triangular sawtooth function. Note that

$$d_\infty(g, h) = \frac{\epsilon}{4} < \frac{\epsilon}{2}.$$

This implies that $g \in B(h, \frac{\epsilon}{2})$. Note that the slope of h is at most L and $\frac{\epsilon}{2}\psi(lx)$ has the slope $\frac{\epsilon}{2}l$. By equation (5.27) we have that the right derivative of g is greater than n. Therefore $g \notin A_n$. In other words, $g \in C[0,1] \setminus A_n$. This implies that $C[0,1] \setminus A_n$ is dense. Since $C[0,1] \setminus A_n$ is open and dense, by the Baire category theorem we get that

$$\bigcap(C[0,1] \setminus A_n) = C[0,1] \setminus \bigcup A_n = C[0,1] \setminus A$$

is dense. $\qquad\square$

Exercises

5.1. First of all, complete whatever is left for you as exercises.

5.2. Show that a complete set in a metric space is closed.

5.3. Can you relax some condition in Cantor's intersection theorem?

5.4. Let (a_n) and (b_n) increasing and decreasing real sequences, respectively, such that $a_n \le b_n$. Show that (a_n) and (b_n) are convergent.

5.5. Let (x_n) be a sequence in a metric space (X, d) such that $\lim_{n \to \infty} d(x_{n+1}, x_n) = 0$. Can you conclude that (x_n) is a Cauchy sequence?

5.6. For a real sequence (x_n), define

$$\lim \sup(x_n) = \inf\{\sup\{x_m \mid m \ge n\} \mid n \in \mathbb{N}\}$$

and

$$\lim \inf(x_n) = \sup\{\inf\{x_m \mid m \ge n\} \mid n \in \mathbb{N}\}.$$

Show that if (x_n) is a bounded real sequence, then there exists a subsequence of (x_n) that converges to $\lim \sup(x_n)$. Also, show that there exists a subsequence of (x_n) that converges to $\lim \inf(x_n)$.

5.7. Show that if (x_n) is a real sequence of positive terms, then

$$\lim \inf \frac{x_{n+1}}{x_n} \le \lim \inf(x_n)^{\frac{1}{n}} \le \lim \sup(x_n)^{\frac{1}{n}} \le \lim \inf \frac{x_{n+1}}{x_n}.$$

5.8. Let V be a normed space. Suppose that W is a closed subspace of V such that W and V/W is complete. Show that V is complete.

5.9. Let $T_m : \ell_p \to \ell_p$ be defined as $T_m((x_n)) = (\frac{x_n}{nm})$, where $m \in \mathbb{N}$. Show that $\|T_m\|$ converges to 0.

5.10. Let $T_m : \ell_p \to \ell_p$ be defined as

$$T_m((x_1, x_2, \dots)) = (0, 0, \dots, x_{m+1}, x_{m+2}, \dots).$$

Show that $T_m((x_n)) \to 0$ but $\|T_m\| \nrightarrow 0$.

5.11. Let V be a separable Banach space. Show that there exists a surjective continuous linear transformation from ℓ_1 to V.

5.12. Let V be a normed space. Let (x_n) be a sequence in V such that $\sum_{n=1}^{\infty} |f(x_n)|$ converges for all linear functionals f. Show that

$$\sup\left\{ \sum_{n=1}^{\infty} |f(x_n)| \;\middle|\; \|f\| \le 1 \right\}$$

is a finite real number.

5.13. Let $T : V \to W$ be a continuous linear transformation that is not surjective, where V is a Banach space, and W is a normed space. Show that if $T(V)$ is dense in W, then it is of first category.

5.14. Let d_1 and d_2 be two Lipschitz equivalent metrics on X. Let $\hat{X}_1$ and $\hat{X}_2$ be completions of (X, d_1) and (X, d_2), respectively. Show that there exist unique continuous maps $f : \hat{X}_1 \to \hat{X}_2$ and $g : \hat{X}_2 \to \hat{X}_1$ such that

$$X \xrightarrow{\ i_1\ } \hat{X}_1$$
$$\searrow{\scriptstyle i_2} \quad \downarrow{\scriptstyle f}$$
$$\hat{X}_2,$$

$$X \xrightarrow{\ i_2\ } \hat{X}_2$$
$$\searrow{\scriptstyle i_1} \quad \downarrow{\scriptstyle g}$$
$$\hat{X}_1,$$

where i_1 and i_2 are isometric embeddings.

5.15. Let A and B be sets in complete metric spaces X and Y, respectively. Let $f : A \to B$ be a homeomorphism. Show that there exist G_δ-sets C and D such that $A \subseteq C \subseteq \mathrm{Cl}\,A$ and $B \subseteq D \subseteq \mathrm{Cl}\,B$ and a homeomorphism $\tilde{f} : C \to D$ such that $\tilde{f}\,|_A = f$.

6 Compact metric spaces

In this chapter, we study an important class of metric spaces called compact metric spaces. We will see that a compact metric space can be embedded in the Hilbert cube as a closed set. Also, we show that every compact metric space is a continuous image of the Cantor set.

6.1 Compactness

Let X be a metric space, and let A be a set in X. Then a family

$$\mathcal{U} = \{U_\alpha \mid \alpha \in \mathcal{A}, \text{where } \mathcal{A} \text{ is an indexing set}\}$$

of open sets of X is called an open cover of A if

$$A \subseteq \bigcup_{\alpha \in \mathcal{A}} U_\alpha.$$

Throughout this chapter, $\mathcal{A}$ will denote an indexing set.

Let $\mathcal{U}$ be an open cover of A. Then a subfamily $\mathcal{V}$ of $\mathcal{U}$ is called a subcover of $\mathcal{U}$ if $\mathcal{V}$ is itself an open cover of A.

Example 6.1.1. Consider the real line $\mathbb{R}$. Let $r > 0$ be a fixed real number. The family

$$\{(x - r, x + r) \mid x \in \mathbb{R}\}$$

is an open cover of $\mathbb{R}$. Note that the family

$$\{(x - r, x + r) \mid x \in \mathbb{Q}\}$$

is a subcover of the above open cover.

Example 6.1.2. The family

$$\mathcal{U} = \{(n, n + 1) \mid n \in \mathbb{Z}\}$$

is not an open cover of $\mathbb{R}$ as $\bigcup_{n \in \mathbb{Z}}(n, n + 1) = \mathbb{R} \setminus \mathbb{Z}$.

Example 6.1.3. The family

$$\mathcal{U} = \{(n - 1, n + 1) \mid n \in \mathbb{Z}\}$$

is an open cover of $\mathbb{R}$. The only subcover of $\mathcal{U}$ is $\mathcal{U}$ itself.

Definition 6.1.4. A set A in a metric space X is called a compact set if every open cover of A has a finite subcover.

https://doi.org/10.1515/9783111636085-006

In other words, if A is compact and

$$\mathcal{U} = \{U_\alpha \mid \alpha \in \mathcal{A}\}$$

is an open cover of A, then it has a subfamily

$$\{U_{\alpha_1}, \ldots, U_{\alpha_n}\}$$

of $\mathcal{U}$ such that $A \subseteq \bigcup_{i=1}^{n} U_{\alpha_i}$. A metric space X is called a compact metric space if X is a compact set of X.

Example 6.1.5. The real line $\mathbb{R}$ is not compact. Consider the open cover

$$\mathcal{U} = \{(-n, n) \mid n \in \mathbb{N}\}$$

of $\mathbb{R}$. On the contrary, suppose that it has a finite subcover

$$\mathcal{V} = \{(-n_1, n_1), \ldots, (-n_l, n_l)\}.$$

Let $k = \max\{n_1, \ldots, n_l\}$. Then

$$\mathbb{R} = \bigcup_{i=1}^{l}(-n_i, n_i) = (-k, k),$$

a contradiction.

Example 6.1.6. Consider the set

$$A = \left\{\frac{1}{n} \,\middle|\, n \in \mathbb{N}\right\} \cup \{0\}.$$

Let

$$\mathcal{U} = \{U_\alpha \mid \alpha \in \mathcal{A}\}$$

be an open cover of A. Then $0 \in U_\beta$ for some $\beta \in \mathcal{A}$. Since U_β is open, there is $\delta > 0$ such that $(-\delta, \delta) \subseteq U_\beta$. By the Archimedean property there is $k \in \mathbb{N}$ such that $\frac{1}{k} < \delta$. Then for all $n \geq k$,

$$\frac{1}{n} \in (-\delta, \delta) \subseteq U_\beta.$$

Let $\frac{1}{i} \in U_{\alpha_i}$ for $1 \leq i \leq k - 1$. Then

$$U_{\alpha_1}, \ldots, U_{\alpha_{k-1}}, U_\beta$$

covers A. This shows that A is compact in $\mathbb{R}$.

Example 6.1.7. Let X be a discrete metric space. Then

$$\{\{x\} \mid x \in X\}$$

is an open cover that does not have a nontrival subcover. We can observe that X is compact if and only if X is finite.

Proposition 6.1.8. *A compact set in a metric space is a closed set.*

Proof. Let A be a compact set in a metric space X. We will prove that $X \setminus A$ is open.

Let $x \in X \setminus A$. Let $a \in A$. Then $a \neq x$. Then there are open balls $B(x, r_a)$ and $B(a, s_a)$ such that

$$B(x, r_a) \cap B(a, s_a) = \emptyset.$$

Note that the family

$$\mathcal{U} = \{B(a, s_a) \mid a \in A\}$$

is an open cover of A. Since A is compact, there is a finite subcover

$$\{B(a_1, s_{a_1}), \dots, B(a_n, s_{a_n})\}$$

of $\mathcal{U}$. Let $U = \bigcup_{i=1}^{n} B(a_i, s_{a_i})$. We can observe that U is open and $A \subseteq U$.

Let $V = \bigcap_{i=1}^{n} B(x, r_{a_i})$. Then V is open, and $x \in V$. We claim that U and V are disjoint. Note that for each $1 \leq i \leq n$,

$$B(x, r_{a_i}) \cap B(a_i, s_{a_i}) = \emptyset.$$

This implies that $B(a_i, s_{a_i}) \cap V = \emptyset$ for each $1 \leq i \leq n$. This shows that $U \cap V = \emptyset$. Therefore

$$x \in V \subseteq X \setminus U \subseteq X \setminus A.$$

This implies that x in an interior point of $X \setminus A$. Since $x \in X \setminus A$ is arbitrary, $X \setminus A$ is open. Hence A is closed in X. $\qquad\square$

Proposition 6.1.9. *A closed set in a compact metric space is a compact set.*

Proof. Let F be a closed set in a compact metric space X. Let

$$\mathcal{U} = \{U_\alpha \mid \alpha \in \mathcal{A}\}$$

be an open cover of F. Consider the set $\mathcal{V} = \mathcal{U} \cup \{X \setminus F\}$. Then $\mathcal{V}$ is an open cover of X. Since X is compact, $\mathcal{V}$ has a finite subcover

$$\mathcal{W} = \{U_{\alpha_1}, \dots, U_{\alpha_n}, X \setminus F\}.$$

Note that

$$F \subseteq \bigcup_{i=1}^{n} U_{\alpha_i}.$$

This shows that F is compact. $\qquad\square$

Proposition 6.1.10. *The continuous image of a compact set is compact.*

Proof. Let f be a continuous map from a metric space X to a metric space Y. Let A be a compact set in X. We claim that $f(A)$ is compact in Y. Let

$$\mathcal{U} = \{U_\alpha \mid \alpha \in \mathcal{A}\}$$

be an open cover of $f(A)$. Since f is continuous, $f^{-1}(U_\alpha)$ is open in X. Since

$$f(A) \subseteq \bigcup_{\alpha \in \mathcal{A}} U_\alpha,$$

we have

$$A \subseteq f^{-1}(f(A))$$
$$\subseteq f^{-1}\left(\bigcup_{\alpha \in \mathcal{A}} U_\alpha \right)$$
$$= \bigcup_{\alpha \in \mathcal{A}} f^{-1}(U_\alpha).$$

This implies that

$$\{f^{-1}(U_\alpha) \mid \alpha \in \mathcal{A}\}$$

is an open cover of A.

Since A is compact, it has a finite subcover

$$\{f^{-1}(U_{\alpha_1}), \ldots, f^{-1}(U_{\alpha_n})\}.$$

Then

$$f(A) \subseteq f\left(\bigcup_{i=1}^{n} f^{-1}(U_{\alpha_i}) \right)$$
$$\subseteq \bigcup_{i=1}^{n} U_{\alpha_i}.$$

This shows that $f(A)$ is compact in Y. $\qquad\square$

Corollary 6.1.11. *Let X and Y be homeomorphic metric spaces, and let X be a compact metric space. Then Y is a compact metric space.*

Corollary 6.1.12. *Let f be a continuous map from a metric space X to a metric space Y. Let X be compact. If F is a closed set in X, then $f(F)$ is a closed set in Y.*

Proposition 6.1.13. *Let X and Y be metric spaces with compact X. Let $f : X \to Y$ be a bijective continuous map. Then f is a homeomorphism.*

Proof. By Corollary 6.1.12 we observe that f is a closed map. Hence f is a homeomorphism. $\qquad\square$

Proposition 6.1.14. *A compact metric space is bounded.*

Proof. Let X be a compact metric space. Let $x \in X$. Then the family

$$\{B(x,n) \mid n \in \mathbb{N}\}$$

is an open cover of X. Since X is compact, it has a finite subcover

$$\{B(x,n_1),\dots,B(x,n_l)\}.$$

Let $k = \max\{n_1,\dots,n_l\}$. Then

$$X = \bigcup_{i=1}^{l} B(x,n_i) = B(x,k).$$

Hence X is bounded. $\qquad\square$

Proposition 6.1.15. *Let A be a compact set in a metric space X, and let $f : A \to \mathbb{R}$ be continuous. Then for all $x \in A$, there exist $a, b \in A$ such that*

$$f(a) \le f(x) \le f(b).$$

Proof. Since A is compact, $f(A)$ is compact in $\mathbb{R}$. By Proposition 6.1.14 it is bounded. Let

$$m = \inf\{f(x) \mid x \in A\} \quad \text{and} \quad M = \sup\{f(x) \mid x \in A\}.$$

We claim that there is $b \in A$ such that $f(b) = M$. On the contrary, suppose that $f(x) < b$ for all $x \in A$. Consider the map $g : A \to \mathbb{R}$ defined by $g(x) = \frac{1}{M-f(x)}$. Note that g is continuous but not a bounded map. This is a contradiction.

Similarly, there is $a \in A$ such that $f(a) = m$. Hence

$$f(a) \le f(x) \le f(b).$$
$\qquad\square$

Let A be a set in a metric space X, and let $\epsilon > 0$. We define the set

$$B(A, \epsilon) = \{z \in X \mid d(z, A) < \epsilon\}.$$

Proposition 6.1.16. *Let A be a compact set in a metric space X, and let U be an open set in X such that $A \subseteq U$. Then there is $\epsilon > 0$ such that $B(A, \epsilon) \subseteq U$.*

Proof. Let $F = X \setminus U$. Define the map $f : X \to \mathbb{R}$ by $f(x) = d(x, F)$. Then f is continuous. Note that $f(x) > 0$ for all $x \in U$. In particular, $f(x) > 0$ for all $x \in A$. By Proposition 6.1.15 there is $a \in A$ such that $f(a) \leq f(x)$ for all $x \in A$. Let $\epsilon = f(a) > 0$. For $x \in A$, we have

$$f(a) \leq f(x) = d(x, F) \leq d(x, y), \quad \text{where } y \in F.$$

This implies that $d(x, y) \geq \epsilon$ for all $x \in A$ and $y \in F$. In other words, if $d(x, y) < \epsilon$ for $x \in A$, then $y \in U$. This implies that

$$B(A, \epsilon) = \{z \in X \mid d(z, A) < \epsilon\} \subseteq U. \qquad \square$$

Theorem 6.1.17 (Heine–Borel theorem). *Let $a, b \in \mathbb{R}$. Then the closed interval $[a, b]$ is compact in $\mathbb{R}$.*

Proof. If $b \leq a$, then $[a, b]$ is at most singleton. Therefore it is compact. Assume that $a < b$. By Corollary 6.1.11 it is sufficient to show that $[0, 1]$ is compact in $\mathbb{R}$. Let

$$\mathcal{U} = \{U_\alpha \mid \alpha \in \mathcal{A}\}$$

be an open cover of $[0, 1]$. Consider the set

$$Y = \{x \in [0, 1] \mid [0, x] \text{ is contained in a union of finitely many members of } \mathcal{U}\}.$$

Note that $0 \in U_\beta$ for some $\beta \in \mathcal{A}$. Then $[0, 0] = \{0\} \subseteq U_\beta$. This implies that $Y \neq \emptyset$. Observe that the set Y is bounded above by 1. By the least upper bound property of $\mathbb{R}$ the supremum of Y exists in $\mathbb{R}$. Let $l = \sup Y$. Clearly, $l \in [0, 1]$. We claim that $l = 1$. This will prove that $[0, 1]$ is compact.

On the contrary, suppose that $l < 1$. Consider an open set $U_\gamma \in \mathcal{U}$ such that $l \in U_\gamma$. Since U_γ is open, there is $\delta > 0$ such that $(l - \delta, l + \delta) \subseteq U_\gamma$ with $l + \delta \in [0, 1]$. Since $l - \delta$ is not an upper bound of Y, there is $y \in Y$ such that $l - \delta < y \leq l$. Let

$$U_{\alpha_1}, \ldots, U_{\alpha_n}$$

be members of $\mathcal{U}$ such that

$$[0, y] \subseteq \bigcup_{i=1}^{n} U_{\alpha_i}.$$

Then

$$[0, l + \delta] = [0, y] \cup [y, l + \delta] \subseteq \bigcup_{i=1}^{n} U_{a_i} \cup U_y.$$

This implies that $l + \delta \in Y$. This is a contradiction since l is the supremum of Y. Hence $l = 1$. $\qquad \square$

Let $a_1, \ldots, a_n, b_1, \ldots, b_n$ be real numbers such that $a_i \leq b_i, 1 \leq i \leq n$. Consider the n-dimensional cuboid

$$C_n = [a_1, b_1] \times \cdots \times [a_n, b_n].$$

We can observe that C_n is closed in $\mathbb{R}^n$.

Theorem 6.1.18. *An n-dimensional cuboid is compact in $\mathbb{R}^n$.*

Proof. The proof is by induction on n. By the Heine–Borel theorem the result is true for $n = 1$. Suppose that every $(n - 1)$-dimensional cuboid is compact in $\mathbb{R}^n$. Let

$$C_n = [a_1, b_1] \times \cdots \times [a_n, b_n].$$

Consider the nth projection $\pi_n : \mathbb{R}^n \to \mathbb{R}$ by $\pi_n(x_1, \ldots, x_n) = x_n$. Let $v \in [a_n, b_n]$. Consider the set

$$C(v) = \{x \in C_n \mid \pi_n(x) = v\}$$
$$= [a_1, b_1] \times [a_{n-1}, b_{n-1}] \times \cdots \times \{v\}.$$

Since $C(v)$ is homeomorphic to C_{n-1}, by induction hypothesis $C(v)$ is compact for all $v \in [a_n, b_n]$.

Let

$$\mathcal{U} = \{U_\alpha \mid \alpha \in \mathcal{A}\}$$

be an open cover of C_n. Note that $\mathcal{U}$ is an open cover of $C(v)$ for all $v \in [a_n, b_n]$. Since $C(v)$ is compact, there is a finite subcover $\mathcal{U}_v$ of $\mathcal{U}$ such that

$$C(v) \subseteq \bigcup_{U \in \mathcal{U}_v} U.$$

Let $V(v) = \bigcup_{U \in \mathcal{U}_v} U$. Then $C(v) \subseteq V(v)$. By Proposition 6.1.16 there is $\epsilon_v > 0$ such that $B(C(v), \epsilon_v) \subseteq V(v)$. Note that for $z \in C_n$ and $y \in C(v)$, we have $|\pi_n(z) - v| \leq d(z, y)$. Then

$$\{z \in C_n \mid |\pi_n(z) - v| < \epsilon_v\} = \{z \in C_n \mid v - \epsilon_v < \pi_n(z) < v + \epsilon_v\}$$
$$\subseteq V(v).$$

Note that $\{(v - \epsilon_v, v + \epsilon_v) \mid v \in [a_n, b_n]\}$ is an open cover of $[a_n, b_n]$. Since $[a_n, b_n]$ is compact, it has a finite subcover

$$\{(v_1 - \epsilon_{v_1}, v_1 + \epsilon_{v_1}), \ldots, (v_k - \epsilon_{v_k}, v_k + \epsilon_{v_k})\}$$

such that

$$[a_n, b_n] \subseteq \bigcup_{i=1}^{k} (v_i - \epsilon_{v_i}, v_i + \epsilon_{v_i}).$$

Note that

$$C_n \subseteq \bigcup_{i=1}^{k} V(v_i).$$

This shows that C_n is compact. $\qquad\qquad\square$

Theorem 6.1.19. *A set A in $\mathbb{R}^n$ is compact if and only if it is closed and bounded.*

Proof. A compact set in a metric space is closed and bounded. Conversely, suppose that A is a closed and bounded set in $\mathbb{R}^n$. Since A is bounded, A is contained in some n-dimensional cuboid C_n. Since C_n is compact and A is closed, A is compact. $\qquad\square$

We have seen that there are several norms on a normed space. We are interested how these norms are related. We have observed that on a normed space the Lipschitz equivalence and topological equivalence coincide. Therefore we will shortly say "equivalent" for "Lipschitz equivalent" or "topological equivalent". Let us focus on the finite-dimensional normed spaces.

Theorem 6.1.20. *Any two norms on a finite-dimensional vector space are equivalent.*

Proof. Let V be a finite-dimensional vector space over $\mathbb{F}$, where $\mathbb{F} = \mathbb{R}$ or $\mathbb{F} = \mathbb{C}$. Let $\| \cdot \|$ be a norm on V. Let $\{v_1, \ldots, v_n\}$ be a basis of V. Then given $x \in V$, $x = x_1 v_1 + \cdots + x_n v_n$, where $x_i \in \mathbb{F}$. Define $\| \cdot \|_2 : V \to \mathbb{R}$ by

$$\|x\|_2 = \left(\sum_{i=1}^{n} |x_i|^2 \right)^{\frac{1}{2}}.$$

We can check that $\| \cdot \|_2$ is a norm on V. We will show that $\| \cdot \|$ is equivalent to $\| \cdot \|_2$. Let $M = \left(\sum_{i=1}^{n} \|v_i\|^2 \right)^{\frac{1}{2}}$. Now

$$\|x\| = \left\| \sum_{i=1}^{n} x_i v_i \right\|$$

$$\leq \sum_{i=1}^{n} |x_i| \|v_i\|$$

$$\leq \left(\sum_{i=1}^{n} |x_i|^2 \right)^{\frac{1}{2}} \left(\sum_{i=1}^{n} \|v_i\|^2 \right)^{\frac{1}{2}} \quad \text{(by the Cauchy–Schwarz inequality)}$$

$$= M \|x\|_2.$$

Let us identify V isometrically with $\mathbb{F}^n$. Define the map $f : V \to \mathbb{R}$ by

$$f(x) = \|x\| = \left\| \sum_{i=1}^{n} x_i v_i \right\|.$$

Using the sequential criterion, we can check that f is continuous with the usual metric on $\mathbb{F}^n$. Let

$$S = \{x \in V \mid \|x\|_2 = 1\}$$
$$= \{(x_1, \ldots, x_n) \in \mathbb{F}^n \mid |x_1|^2 + \cdots + |x_n|^2 = 1\}.$$

If $\mathbb{F} = \mathbb{R}$, then S can be identified with the unit sphere in $\mathbb{R}^n$. If $\mathbb{F} = \mathbb{C}$, then S can be identified with the unit sphere in $\mathbb{R}^{2n}$. By Theorem 6.1.19, S is compact. By Proposition 6.1.15 there is $a = (a_1, \ldots, a_n) \in S$ such that $f(a) \leq f(x)$ for all $x \in S$. Let $m = f(a)$. Suppose that $m = 0$. Then

$$f(a) = \left\| \sum a_i v_i \right\| = 0,$$

and thus $\sum a_i v_i = 0$. This implies that $a_i = 0$ for all i, a contradiction. Hence $m > 0$.

Let $y \in V$ be such that $y \neq 0$. Then $\|\frac{y}{\|y\|_2}\|_2 = 1$. This implies that $\frac{y}{\|y\|_2} \in S$. Then

$$m = f(a) \leq f\left(\frac{y}{\|y\|_2} \right) = \left\| \frac{y}{\|y\|_2} \right\|.$$

Therefore

$$m\|y\|_2 \leq \|y\|. \tag{6.1}$$

If $y = 0$, then inequality (6.1) is trivially satisfied. Therefore $m\|x\|_2 \leq \|x\|$ for all $x \in \mathbb{F}^n$. Thus we have

$$m\|x\|_2 \leq \|x\| \leq M\|x\|_2.$$

Hence $\|\cdot\|$ is equivalent to $\|\cdot\|_2$. Since equivalence of norms is an equivalence relation, any two norms on V are equivalent. □

Now there is a natural question whether there are inequivalent norms on an infinite-dimensional normed space. If there are such norms, then how many inequivalent norms are there. The following result answers this question. Its proof uses some cardinal arithmetic. If the reader is not aware of this, then he or she may leave the proof with the understanding that there are rather many inequivalent norms on an infinite-dimensional normed space.

Theorem 6.1.21 (Miyeon Kwon). *If V is infinite dimensional, then the number of inequivalent norms on V is $2^{\dim(V)}$.*

Proof. Let $\{v_a \mid a \in \mathcal{A}\}$ be a basis of V. Since $\mathcal{A}$ is infinite, the set $\mathcal{A} \times \mathbb{N}$ is in bijection with $\mathcal{A}$. Without loss of generality, we can index the elements of a basis with $\mathcal{A} \times \mathbb{N}$.

Suppose that $\{v_{an} \mid (a,n) \in \mathcal{A} \times \mathbb{N}\}$ is a basis of V. Let $x \in V$. Then

$$x = \sum_{(a,n) \in \mathcal{A} \times \mathbb{N}} x_{an} v_{an},$$

where the sum is finite. For each subset $\mathcal{I}$ of $\mathcal{A}$, define the norm on V by

$$\|x\|_{\mathcal{I}} = \left\| \sum_{(a,n) \in \mathcal{A} \times \mathbb{N}} x_{an} v_{an} \right\|_{\mathcal{I}}$$

$$= \sum_{(a,n) \in \mathcal{I} \times \mathbb{N}} \frac{1}{n} |x_{an}| + \sum_{(a,n) \in (\mathcal{A} \setminus \mathcal{I}) \times \mathbb{N}} |x_{an}|.$$

Suppose that $\mathcal{I}$ and $\mathcal{J}$ are distinct subsets of $\mathcal{A}$. Let $a \in \mathcal{I} \setminus \mathcal{J}$. Then we can check that $\|v_{an}\|_{\mathcal{I}} \to 0$ and $\|v_{an}\|_{\mathcal{J}} \to 1$ as $n \to \infty$. This implies that $\|\ \|_{\mathcal{I}}$ is not equivalent to $\|\ \|_{\mathcal{J}}$. Therefore the number of inequivalent norms on V is at least the number of subsets of $\mathcal{A}$, that is, $2^{\dim(V)}$.

First, suppose that $\mathbb{F} = \mathbb{R}$. For a basis $\mathcal{B}$ of V, consider the vector space T over the field $\mathbb{Q}$ of rationals. Then

$$T = \{x_1 v_1 + \cdots + x_n v_n \mid x_i \in \mathbb{Q}, v_i \in \mathcal{B}, n \in \mathbb{N}\}.$$

Since $\mathbb{Q}$ is dense in the real line, each real number is a limit of some sequence in $\mathbb{Q}$. Also, since a norm is a continuous map, each norm on V is completely determined by its values in T. This implies that the number of norms on V is at most the number of maps from T to $\mathbb{R}$. Let $\mathbf{c}$ be the cardinality of $\mathbb{R}$. Then the number of maps from T to $\mathbb{R}$ is

$$\mathbf{c}^{|T|} = \left(2^{\aleph_0}\right)^{|T|}$$
$$= \left(2^{\aleph_0}\right)^{|\mathcal{B}|} \quad \text{as } |\mathcal{B}| = |T|$$
$$= 2^{\aleph_0 |\mathcal{B}|}$$
$$= 2^{|\mathcal{B}|}$$
$$= 2^{\dim(V)}.$$

Thus the number of inequivalent norms on V is $2^{\dim(V)}$.

If $\mathbb{F} = \mathbb{C}$, then we consider T as a vector space over the field $\mathbb{Q} + i\mathbb{Q}$, and the rest of the proof is similar. $\square$

We observed above the first deviation from a finite-dimensional normed space to an infinite-dimensional normed space in terms of equivalence of norms. Let us see the second deviation. Suppose that V is a finite-dimensional normed space over $\mathbb{F}$. We can observe that the unit sphere in V is compact. We will show that the compactness of the unit sphere in V implies that V is finite dimensional. First, we observe the following:

Theorem 6.1.22. *Let V be a normed space, and let W be a proper closed subspace of V. Let a be a real number such that $0 < a < 1$. Then there is $x_a \in V$ such that $\|x_a\| = 1$ and $\|x_a - y\| > a$ for all $y \in W$.*

Proof. Choose a point $x \in V \setminus W$. Since W is closed,

$$d = d(x, W) = \inf\{\|x - w\| \mid w \in W\} > 0.$$

Since $0 < a < 1$, $d < \frac{d}{a}$. Then there is $w \in W$ such that $\|x - w\| < \frac{d}{a}$. Let $x_a = \frac{x-w}{\|x-w\|}$. Clearly, $\|x_a\| = 1$. For any $y \in W$, we have

$$\|x_a - y\| = \left\| \frac{x - w}{\|x - w\|} - y \right\|$$
$$= \frac{1}{\|x - w\|} \|x - (w + \|x - w\|y)\|$$
$$> \frac{a}{d} d$$
$$= a. \qquad \square$$

Now we prove the following characterization of a finite-dimensional normed space. To prove it, we need that if a set A is compact in X, then every sequence in A has a convergent subsequence. We will prove this fact later.

Theorem 6.1.23. *A normed space V is finite dimensional if and only if the unit sphere $S = \{x \in V \mid \|x\| = 1\}$ is compact.*

Proof. If V is finite dimensional, then we can easily prove that S is compact in V. To prove the converse, suppose that V is an infinite-dimensional normed space.

Choose $x_1 \in S$. Since V is infinite dimensional, the subspace $\langle x_1 \rangle$ generated by x_1 is not V. Since $\langle x_1 \rangle$ is finite dimensional, $\langle x_1 \rangle$ is closed in V. Let a be a fixed real number such that $0 < a < 1$. By Theorem 6.1.22 there is $x_2 \in S$ such that $\|x_2 - y\| > a$ for all $y \in \langle x_1 \rangle$. Again, $\langle x_1, x_2 \rangle \neq V$. By Theorem 6.1.22 there is $x_3 \in S$ such that $\|x_3 - y\| > a$ for all $y \in \langle x_1, x_2 \rangle$. In this way, we get a sequence (x_n) in S such that $\|x_n - x_m\| > a$ for $n \neq m$. This sequence does not have a convergent subsequence. This shows that S is not compact. $\qquad \square$

By the same argument and the same sequence (x_n) we can prove that a normed space V is finite dimensional if and only if the closed unit ball $D = \{x \in V \mid \|x\| \leq 1\}$ is compact.

Let us see one more characterization of a finite-dimensional normed space.

Theorem 6.1.24. *A normed space V is finite dimensional if and only if every linear functional $f : V \to \mathbb{F}$ is continuous.*

Proof. Suppose that V is a finite-dimensional normed space with $\dim V = n$. Let $\{v_1, \ldots, v_n\}$ be a basis of V. Then every $x \in V$ can be uniquely written as $x = \sum_{i=1}^{n} x_i v_i$, where $x_i \in \mathbb{F}$. Since V is finite dimensional, it is sufficient to consider the norm $\|\cdot\|_1$ on V defined by $\|x\|_1 = \sum_{i=1}^{n} |x_i|$.

Let $f : V \to \mathbb{F}$ be a linear functional. Let $M = \max\{|f(v_1)|, \ldots, |f(v_n)|\}$. Then

$$
\begin{aligned}
|f(x)| &= \left| \sum_{i=1}^{n} x_i f(v_i) \right| \\
&\leq \sum_{i=1}^{n} |x_i| |f(v_i)| \\
&\leq M \sum_{i=1}^{n} |x_i| \\
&= M \|x\|_1.
\end{aligned}
$$

Therefore f is continuous.

To prove the converse, suppose that V is an infinite-dimensional normed space. Consider a linearly independent set $\{v_i \mid i \in \mathbb{N}\}$ in V. Without loss of generality, we can assume that $\|v_i\| = 1$ for all $i \in \mathbb{N}$. Extend this linearly independent set to a basis

$$
\mathcal{B} = \{v_i \mid i \in \mathbb{N}\} \cup \{w_\alpha \mid \alpha \in \mathcal{A}\}
$$

of V, where $\mathbb{N} \cap \mathcal{A} = \emptyset$. Define the linear functional $f : V \to \mathbb{R}$ by $f(v_i) = i$ and $f(w_\alpha) = 0$ for all $i \in \mathbb{N}$ and $\alpha \in \mathcal{A}$. We claim that f is discontinuous. On the contrary, suppose that f is continuous. Then $|f(v_i)| \leq \|f\| \|v_i\|$. This implies that $\|f\| \geq i$ for all $i \in \mathbb{N}$, a contradiction. $\square$

By Theorems 6.1.23, 6.1.24, and 3.1.34 we get the following:

Theorem 6.1.25. *Let V be a normed space V. Then following statements are equivalent:*
(i)   *V is finite dimensional;*
(ii) *The unit sphere $S = \{x \in V \mid \|x\| = 1\}$ is compact;*
(iii) *The closed unit ball $D = \{x \in V \mid \|x\| \leq 1\}$ is compact;*
(iv) *Every linear functional $f : V \to \mathbb{F}$ is continuous;*
(v)   *Every subspace of V is closed.*

6.2 Equivalence of compactness

In this section, we study some properties of a metric space, which are equivalent to the compactness.

Definition 6.2.1. A metric space X is said to satisfy the Bolzano–Weierstrass property if every infinite set in X has a limit point.

Proposition 6.2.2. *Let X be a compact metric space. Then X satisfies the Bolzano–Weierstrass property.*

Proof. On the contrary, suppose that X does not satisfy the Bolzano–Weierstrass property. Let A be an infinite set in X that has no limit point in X. Then A is closed. Therefore A is compact.

Let $a \in A$. Since a is not a limit point of A, there is an open ball $B(a, r_a)$ that does not contain any other point of A. Note that the family

$$\{B(a, r_a) \mid a \in A\}$$

is an open cover of A. Since A is compact, it has a finite subcover

$$\{B(a_1, r_{a_1}), \ldots, B(a_n, r_{a_n})\}$$

such that

$$A \subseteq \bigcup_{i=1}^{n} B(a_i, r_{a_i}).$$

Since each open ball $B(a_i, r_{a_i})$ contains only one point of A, namely a_i, A is a finite set, a contradiction. $\square$

Definition 6.2.3. A metric space X is called sequentially compact if every sequence in X has a convergent subsequence.

Proposition 6.2.4. *Let X be a metric space that satisfies the Bolzano–Weierstrass property. Then X is sequentially compact.*

Proof. Consider a sequence (x_n) in X. Let A be the set of terms of the sequence (x_n).

First, suppose that A is a finite set. Then there is term of the sequence (x_n), say x, which is repeated infinitely many times. Then the sequence (x_n) has a constant subsequence (x). Clearly, the subsequence (x) converges to x.

Now suppose that A is an infinite set. Since X satisfies the Bolzano–Weierstrass property, A has a limit point in X. Let $x \in X$ be a limit point of A. Choose a term x_{n_1} in the open ball $B(x, 1)$. Next, choose a term x_{n_2} in the open ball $B(x, \frac{1}{2})$, where $n_2 > n_1$. In this way, we obtain a subsequence (x_{n_k}) of the sequence (x_n), where $x_{n_k} \in B(x, \frac{1}{k})$. We can easily observe that $x_{n_k} \to x$. $\square$

Proposition 6.2.5. *Let X be a sequentially compact metric space. Let $\mathcal{U}$ be an open cover of X by open balls. Then there is $n \in \mathbb{N}$ such that every open ball of diameter less than $\frac{1}{n}$ is contained in some member of $\mathcal{U}$.*

Proof. On the contrary, suppose for each $n \in \mathbb{N}$, there is an open ball $B(x_n, \frac{1}{n})$ that is not contained in any member of $\mathcal{U}$. This gives a sequence (x_n) in X. Since X is sequentially compact, (x_n) has a convergent subsequence (x_{n_i}). Suppose $x_{n_i} \to x$ in X.

Since $\mathcal{U}$ covers X, there is an open ball $B(y,r)$ in $\mathcal{U}$ containing x. Let

$$s = \min\{d(x,y), r - d(x,y)\}.$$

Then $B(x,s) \subseteq B(y,r)$. Since $x_{n_i} \to x$, there is $k \in \mathbb{N}$ such that

$$j \geq k \Rightarrow d(x_{n_j}, x) < \frac{s}{2}.$$

Choose $n_i \geq n_k$ such that $\frac{1}{n_i} < \frac{s}{2}$. Since $n_i \geq n_k$, $d(x_{n_i}, x) < \frac{s}{2}$. We claim that $B(x_{n_i}, \frac{1}{n_i}) \subseteq B(x,s)$. For this, let $z \in B(x_{n_i}, \frac{1}{n_i})$. Then $d(z, x_{n_i}) < \frac{1}{n_i}$. Now

$$d(z,x) \leq d(z, x_{n_i}) + d(x_{n_i}, x)$$
$$< \frac{1}{n_i} + \frac{s}{2}$$
$$< \frac{s}{2} + \frac{s}{2}$$
$$= s.$$

This implies that $z \in B(x,s)$, and thus

$$B\left(x_{n_i}, \frac{1}{n_i}\right) \subseteq B(x,s) \subseteq B(y,r).$$

This is a contradiction. □

Let X be a sequentially compact metric space, and let $\mathcal{U}$ be an open cover of X. By Proposition 6.2.5 there is a positive real number δ such that any set in X of diameter less than δ is contained in some member of $\mathcal{U}$. Such a positive real number δ is called a Lebesgue number for the open cover $\mathcal{U}$. Note that a Lebesgue number is not unique for an open cover $\mathcal{U}$. Any positive real number less than δ is also a Lebesgue number for $\mathcal{U}$.

Proposition 6.2.6. *Every continuous map from a compact metric space is uniformly continuous.*

Proof. Let X and Y be metric spaces with compact X. Let $f : X \to Y$ be a continuous map. Let $\epsilon > 0$. Then the family

$$\mathcal{U} = \left\{ B\left(y, \frac{\epsilon}{2}\right) \mid y \in Y \right\}$$

is an open cover of Y. Since f is continuous, the family

$$\mathcal{V} = \left\{ f^{-1}\left(B\left(y, \frac{\epsilon}{2}\right)\right) \mid y \in Y \right\}$$

is an open cover of X. Since X is compact, X is sequentially compact. Let $\delta > 0$ be a Lebesgue number for $\mathcal{V}$. Let $x_1, x_2 \in X$ be such that $d(x_1, x_2) < \delta$. In other words, the diameter of the set $\{x_1, x_2\}$ is less than δ. Choose a member $f^{-1}(B(y, \frac{\epsilon}{2}))$ of $\mathcal{V}$ containing x_1 and x_2. Then $f(x_1), f(x_2) \in B(y, \frac{\epsilon}{2})$. Now

$$d(f(x_1), f(x_2)) \le d(f(x_1), y) + d(y, f(x_2)) < \epsilon.$$

Hence f is uniformly continuous. $\qquad\square$

Remark 6.2.7. If we carefully observe the proof of Proposition 6.2.6, then we can see that if a metric space X in which a Lebesgue number exists for every open cover of X, then every continuous map on X is uniformly continuous.

Definition 6.2.8. A set A in a metric space X is called totally bounded if for each $\epsilon > 0$, there are finitely many points $x_1, \dots, x_n$ such that

$$A \subseteq \bigcup_{i=1}^{n} B(x_i, \epsilon).$$

A metric space X is called totally bounded if X is a totally bounded set in X.

Example 6.2.9. Consider the open interval $(0, 1)$ in the real line $\mathbb{R}$. Let $\epsilon > 0$. If $\epsilon \ge 1$, then $(0, 1) \subseteq B(x, \epsilon)$, where $x \in (0, 1)$. Let $0 < \epsilon < 1$. Then there exists $n \in \mathbb{N}$ such that $\frac{1}{n+1} < \epsilon$. Let $x_i = \frac{1}{n+1}, i = 1, \dots, n$. We can observe that

$$(0, 1) \subseteq \bigcup_{i=1}^{n} B(x_i, \epsilon).$$

Therefore $(0, 1)$ is totally bounded.

Proposition 6.2.10. *A totally bounded set in a metric space is bounded.*

Proof. Let A be a totally bounded set in a metric space X. Then for $\epsilon = 1$, there is a finite set $Y = \{x_1, \dots, x_n\}$ such that

$$A \subseteq \bigcup_{i=1}^{n} B(x_i, 1).$$

Let $x, y \in A$. Then $x \in B(x_i, 1)$ and $y \in B(x_j, 1)$ for some $x_i, x_j \in Y$. Now

$$d(x, y) \le d(x, x_i) + d(x_i, x_j) + d(x_j, y)$$
$$< 2 + \mathrm{diam}(Y).$$

This shows that A is bounded. $\qquad\square$

A bounded set need not be totally bounded. Consider the set of real numbers $\mathbb{R}$ with the metric

$$d(x,y) = \min\{1, |x-y|\}.$$

Then $\mathbb{R}$ is bounded but not totally bounded. Also, an infinite set with the discrete metric is bounded but not totally bounded, as any open ball of radius less than 1 is a singleton set.

Proposition 6.2.11. *A sequentially compact metric space is totally bounded.*

Proof. Let X be a sequentially compact metric space. On the contrary, suppose that X is not totally bounded. Then there is $\epsilon > 0$ for which there are no finitely many points $x_1, \ldots, x_n$ such that

$$\bigcup_{i=1}^{n} B(x_i, \epsilon) \neq X.$$

Choose a point $x_1 \in X$ and consider the open ball $B(x_1, \epsilon)$. If $B(x_1, \epsilon) = X$, then X is totally bounded, a contradiction. Thus $B(x_1, \epsilon) \neq X$. Choose $x_2 \in X \setminus B(x_1, \epsilon)$. Consider $B(x_1, \epsilon) \cup B(x_2, \epsilon)$. Since X is not totally bounded, we inductively get a sequence (x_n) such that $d(x_n, x_m) \geq \epsilon$ for $n \neq m$. We can observe that this sequence (x_n) does not have a convergent subsequence. This is a contradiction. ☐

Proposition 6.2.12. *A sequentially compact metric space is compact.*

Proof. Let X be a sequentially compact metric space. Let $\mathcal{U}$ be an open cover of X. Let $\delta >$ be a Lebesgue number for the open cover $\mathcal{U}$. Since a sequentially compact metric space is totally bounded, for $\frac{\delta}{3}$, there are finitely many points $x_1, \ldots, x_n$ such that

$$X = \bigcup_{i=1}^{n} B\left(x_i, \frac{\delta}{3}\right).$$

Note that $\mathrm{diam}(B(x_i, \frac{\delta}{3})) \leq \frac{2\delta}{3} < \delta$ for all $i = 1, \ldots, n$. Let $U_i \in \mathcal{U}$ be such that $B(x_i, \frac{\delta}{3}) \subseteq U_i$ for all $i = 1, \ldots, n$. This implies that

$$X = \bigcup_{i=1}^{n} U_i.$$

This shows that X is compact. ☐

Proposition 6.2.13. *A compact metric space is complete.*

Proof. Let X be a compact metric space. Let (x_n) be a Cauchy sequence in X. Since a compact metric space is sequentially compact, (x_n) has a convergent subsequence. By Proposition 5.2.21 X is complete. ☐

The real line $\mathbb{R}$ is complete but not compact. It is natural to ask when a complete metric space is compact. We have the following:

Proposition 6.2.14. *A metric space X is compact if and only if X is complete and totally bounded.*

Proof. We have observed that a sequentially compact metric space is totally bounded and a compact metric space is complete. Therefore a compact metric space is complete and totally bounded. Conversely, suppose that X is complete and totally bounded. We will show that X satisfies the Bolzano–Weierstrass property.

Let A be an infinite set in X. Since X is totally bounded, for each $n \in \mathbb{N}$, there is a finite set Y_n in X such that

$$X = \bigcup_{x \in Y_n} B\left(x, \frac{1}{n}\right).$$

Then for $n = 1$, there is a point $x_1 \in Y_1$ such that $B(x, 1)$ contains infinitely many points of A. In other words, $A \cap B(x, 1)$ is infinite. Since

$$A \cap B(x_1, 1) \subseteq \bigcup_{x \in Y_2} B\left(x, \frac{1}{2}\right),$$

there is a point $x_2 \in Y_2$ such that

$$A \cap B(x_1, 1) \cap B\left(x_2, \frac{1}{2}\right)$$

is infinite. Inductively, we get $x_n \in Y_n$ such that

$$A \cap B(x_1, 1) \cap \cdots \cap B\left(x_n, \frac{1}{n}\right)$$

is infinite. Let $\epsilon > 0$. Then there is $k \in \mathbb{N}$ such that $\frac{2}{k} < \epsilon$. Let $m, n \geq k$ and $a \in A \cap B(x_m, \frac{1}{m}) \cap B(x_n, \frac{1}{n})$. Then

$$d(x_m, x_n) \leq d(x_m, a) + d(a, x_n)$$
$$< \frac{1}{m} + \frac{1}{n}$$
$$\leq \frac{2}{k}$$
$$< \epsilon.$$

This implies that (x_n) is a Cauchy sequence. Since X is complete, $x_n \to x$ in X. Therefore x is a limit point of A. This shows that X satisfies the Bolzano–Weierstrass property. □

We have observed that every continuous real-valued map from a compact metric space is bounded. We are interested in a metric space with every continuous real-valued map bounded.

Definition 6.2.15. A metric space X is called pseudocompact if every continuous real-valued map from X is bounded.

Proposition 6.2.16. *Let X be a pseudocompact metric space. Then a Lebesgue number exists for every open cover of X.*

Proof. Let $\mathcal{U}$ be an open cover of X. If $X \in \mathcal{U}$, then we are done. Suppose that $X \notin \mathcal{U}$. Choose a point $a \in X$. Note that $d(x,a) = d(x,\{a\})$ for all $x \in X$. By Example 3.1.30 the map $f : X \to \mathbb{R}$ defined by $f(x) = d(x,a)$ is continuous. By assumption there is $M > 0$ such that $f(x) \le \frac{M}{2}$ for all $x \in X$. Let $x,y \in X$. Then

$$d(x,y) \le d(x,a) + d(a,y)$$
$$\le \frac{M}{2} + \frac{M}{2}$$
$$= M.$$

We define the map $\phi : X \to \mathbb{R}$ by

$$\phi(x) = \sup\{d(x, X \setminus U) \mid U \in \mathcal{U}\}.$$

Note that

$$d(x, X \setminus U) = \inf\{d(x,y) \mid y \in X \setminus U\} \le M.$$

This implies that $\phi(x) \le M$ for all $x \in X$. For all $x,y \in X$ and $U \in \mathcal{U}$, we have

$$\left|d(x, X \setminus U) - d(y, X \setminus U)\right| \le d(x,y).$$

Then

$$d(x, X \setminus U) \le d(y, X \setminus U) + d(x,y)$$
$$\le \phi(y) + d(x,y).$$

This implies that

$$\phi(x) \le \phi(y) + d(x,y).$$

Therefore

$$\phi(x) - \phi(y) \le d(x,y).$$

We can similarly show that

$$\phi(y) - \phi(x) \leq d(x,y).$$

Hence

$$|\phi(x) - \phi(y)| \leq d(x,y).$$

This shows that ϕ is a continuous map. Note that $X \setminus U \neq \emptyset$ for all $U \in \mathcal{U}$. Let $x \in X$. Then $x \in V$ for some $V \in \mathcal{U}$. Then

$$\phi(x) \geq d(x, X \setminus V) > 0.$$

Then the map $\psi : X \to \mathbb{R}$ defined by $\psi(x) = \frac{1}{\phi(x)}$ is continuous. By assumption there is $m > 0$ such that $\psi(x) < m$ for all $x \in X$ or, in other words, $\frac{1}{m} < \phi(x)$ for all $x \in X$. We claim that any δ such that $0 < \delta < \frac{1}{m}$ is a Lebesgue number.

Let $x \in X$. Since $0 < \delta < \frac{1}{m}$, there is $U \in \mathcal{U}$ such that $d(x, X \setminus U) > \delta$. Let $z \in B(x, \delta)$. Then $d(z,x) < \delta$. This implies that $z \in U$. Since $z \in B(x,\delta)$ is arbitrary, $B(x,\delta) \subseteq U$. This shows that δ is a Lebesgue number. $\qquad\square$

Proposition 6.2.17. *A pseudocompact metric space is totally bounded.*

Proof. Let X be a pseudocompact metric space. On the contrary, suppose that X is not totally bounded. Then there is $\epsilon > 0$ such that X cannot be written as a union of finitely many balls of radius ϵ.

Let $x_1 \in X$ and consider the open ball $B(x_1, \epsilon)$. Since X is not totally bounded, $B(x_1, \epsilon) \neq X$. Choose $x_2 \in X \setminus B(x_1, \epsilon)$. Again,

$$X \neq B(x_1, \epsilon) \cup B(x_2, \epsilon).$$

Inductively, we get a sequence (x_n) in X such that $d(x_m, x_n) \geq \epsilon$ for $m \neq n$. Let $A = \{x_n \mid n \in \mathbb{N}\}$. Note that A is closed in X and the subspace metric on A is topologically equivalent to the discrete metric on A. Then the map $f : A \to \mathbb{R}$ defined by $f(x) = n\epsilon$ is continuous. By Tietze's extension theorem, f can be extended to a continuous map $\tilde{f} : X \to \mathbb{R}$. Observe that $\tilde{f}$ is unbounded, a contradiction. $\qquad\square$

Proposition 6.2.18. *A pseudocompact metric space is compact.*

Proof. Let X be a pseudocompact metric space. Let $\mathcal{U}$ be an open cover of X. Let $\delta > 0$ be a Lebesgue number for $\mathcal{U}$. Since X is totally bounded, there are finitely many points $x_1, \ldots, x_n$ of X such that

$$X = B(x_1, \delta) \cup \cdots \cup B(x_n, \delta).$$

Since $\delta > 0$ is a Lebesgue number, there is an open set $U_i \in \mathcal{U}$ such that $B(x_i, \delta) \subseteq U_i$ for all $1 \leq i \leq n$. Then

$$X = U_1 \cup \cdots \cup U_n.$$

This shows that X is compact. □

Definition 6.2.19. A metric space X is called countably compact if every countable open cover has a finite subcover.

Clearly, a compact metric space is countably compact.

Proposition 6.2.20. *A countably compact metric space is compact.*

Proof. Let X be a countably compact metric space. We will show that X satisfies the Bolzano–Weierstrass property. On the contrary, suppose that there is an infinite set Y in X that has no limit point. Let A be a countably infinite subset of Y. Then A does not have limit point in X. This implies that for each $x \in X$, there is an open ball $B(x, r)$ such that

$$B(x, r) \cap (A \setminus \{x\}) = \emptyset.$$

Let $a \in A$. Then for every $x \in X \setminus (A \setminus \{a\})$, there is an open ball $B(x, r)$ such that

$$B(x, r) \cap (A \setminus \{a\}) = \emptyset.$$

This implies that $B(x, r) \subseteq X \setminus (A \setminus \{a\})$. Therefore $U_a = X \setminus (A \setminus \{a\})$ is an open set in X. Note that for each $x \in X$, there is an open ball around x contained in U_a. This implies that

$$\mathcal{U} = \{U_a \mid a \in A\}$$

is an open cover of X. Since X is countably infinite, it has a finite subcover

$$\{U_{a_1}, \ldots, U_{a_n}\}.$$

Note that $U_a \cap A = \{a\}$ and

$$A \subseteq U_{a_1} \cup \cdots \cup U_{a_n}.$$

This implies that A is finite, a contradiction. □

Thus for a metric space, the compactness, Bolzano–Weierstrass property, sequential compactness, pseudocompactness, and countable compactness are equivalent.

6.3 Hilbert cube

The aim of this section is to show that a compact metric space can be homeomorphically embedded in the Hilbert cube as a closed set.

Let $I = [0,1]$ with the metric of the real line. Consider the product $I^{\mathbb{N}}$ with the metric

$$d(x,y) = \sum_{n=1}^{\infty} \frac{|x_n - y_n|}{2^n}$$

for $x = (x_n)$ and $y = (y_n)$. The metric space $(I^{\mathbb{N}}, d)$ is called the Hilbert cube.

Let $\mathcal{S}$ be a collection of open sets in a metric space X such that for every $x \in U$, where U is open in X, there are finitely many members $S_1, \ldots, S_n$ in $\mathcal{S}$ such that

$$x \in \bigcap_{i=1}^{n} S_i \subseteq U.$$

Remark 6.3.1. Such a collection $\mathcal{S}$ is called a subbase. We may be curious that there may be also a notion of a base. It is there, but it is deliberately not defined, although we have used this notion in this book.

Theorem 6.3.2 (Alexander subbase theorem). *Let $\mathcal{S}$ be a subbase in a metric space X. If for every open cover $\mathcal{U} \subseteq \mathcal{S}$, there is a finite subcover, then X is compact.*

Proof. On the contrary, suppose that X is not compact. Let $\mathcal{V}$ be an open cover of X that has no finite subcover. Let $\mathcal{V}$ be maximal with respect to this property. Then $\mathcal{S} \cap \mathcal{V}$ cannot cover X, since otherwise by our assumption $\mathcal{S} \cap \mathcal{V} \subseteq \mathcal{S}$ has a finite subcover. This is a contradiction as $\mathcal{S} \cap \mathcal{V} \subseteq \mathcal{V}$.

Let W be the union of members of $\mathcal{S} \cap \mathcal{V}$. Let $x \in X \setminus W$. Then there is an open set $U \in \mathcal{V}$ such that $x \in U$. Since $\mathcal{S}$ is a subbase, there are finitely many members $S_1, \ldots, S_n$ of $\mathcal{S}$ such that

$$x \in \bigcap_{i=1}^{n} S_i \subseteq U.$$

We can observe that $S_i \notin \mathcal{S} \cap \mathcal{V}$; otherwise, if x is in some S_i, then $x \in W$. By the maximality of $\mathcal{V}$ the open cover $\mathcal{V} \cup \{S_i\}$ has finite subcover for each $1 \leq i \leq n$, say

$$\{V_1^i, \ldots, V_{m_i}^i\} \cup \{S_i\}.$$

This implies that for each $1 \leq i \leq n$, $X \setminus \{S_i\}$ is covered by finitely many members of $\mathcal{V}$. Therefore $X \setminus \cap S_i$ and consequently $X \setminus U$ are covered by finitely many members of $\mathcal{V}$. Since $U \in \mathcal{V}$, X is covered by finitely many members of $\mathcal{V}$. This is a contradiction. □

Theorem 6.3.3. *The Hilbert cube is compact.*

Proof. By Proposition 4.1.6, given $x \in V$ with V open in the Hilbert cube, there is an open set

$$G = U_1 \times \cdots \times U_n \times I \times I \times \cdots$$

such that

$$x \in G \subseteq V.$$

For $k \in I$, let $I^k = I \times \cdots \times I$ (k times). For $k \in \mathbb{N}$, $a \in [0, 1)$, and $b \in (0, 1]$, denote

$$L_k(a) = I^{k-1} \times (a, 1] \times I \times I \times \cdots$$

and

$$U_k(b) = I^{k-1} \times [0, b) \times I \times I \times \cdots.$$

Let $\mathcal{S}$ be the set of all such $L_k(a)$ and $U_l(b)$. Note that $\mathcal{S}$ is a subbase. Let $\mathcal{U}$ be an open cover contained in $\mathcal{S}$. For each $k \in \mathbb{N}$, let

$$a_k = \inf\{a \in [0, 1) \mid L_k(a) \in \mathcal{U}\}$$

and

$$b_k = \inf\{b \in (0, 1] \mid U_k(b) \in \mathcal{U}\}.$$

Now we claim that there is $m \in \mathbb{N}$ such that $a_m < b_m$. On the contrary, suppose that $a_n \geq b_n$ for all $n \in \mathbb{N}$. Then

$$x = (a_1, a_2, \dots) \notin \bigcup_{U \in \mathcal{U}} U.$$

This is a contradiction. Therefore $a_m < b_m$ for some $m \in \mathbb{N}$. Then there are $a \in [0, 1)$ and $b \in (0, 1]$ with $a_m \leq a < b \leq b_m$ such that $L_m(a), U_m(b) \in \mathcal{U}$. We can observe that $\{L_m(a), U_m(b)\}$ is a finite subcover of $\mathcal{U}$. By the Alexander subbase theorem, $I^{\mathbb{N}}$ is compact. □

Proposition 6.3.4. *Let* $\mathbf{H} = \{(x_n) \in \ell_2 \mid 0 \leq x_n \leq \frac{1}{n}\}$. *Then* $\mathbf{H}$ *as a metric subspace of* ℓ_2 *is homeomorphic to the Hilbert cube.*

Proof. Let d be the metric on the Hilbert cube. Define the map $f : I^{\mathbb{N}} \to \mathbf{H}$ by $f((x_n)) = (\frac{x_n}{n})$. We can check that f is bijective. Since the Hilbert cube is compact, to prove that f is a homeomorphism, it is sufficient to prove that f is continuous.

Let $\epsilon > 0$. Since $\sum_{n=1}^{\infty} \frac{1}{n^2}$ is convergent, there is $k \in \mathbb{N}$ such that

$$\sum_{n=k+1}^{\infty} \frac{1}{n^2} < \frac{\epsilon^2}{2}.$$

Let δ be a positive real number defined by the equation

$$\sum_{n=1}^{k} \left(\frac{2^n}{n}\right)^2 \delta^2 = \frac{\epsilon^2}{2}.$$

Let $x = (x_n), a = (a_n) \in I^{\mathbb{N}}$ be such that

$$d(x, a) = \sum_{n=1}^{\infty} \frac{|x_n - a_n|}{2^n} < \delta.$$

Then for each $n \in \mathbb{N}$, we have

$$\frac{|x_n - a_n|}{2^n} \leq \sum_{n=1}^{k} \frac{|x_n - a_n|}{2^n}$$

$$\leq \sum_{n=1}^{\infty} \frac{|x_n - a_n|}{2^n}$$

$$< \delta.$$

Now

$$\|f(x) - f(a)\|_2^2 = \sum_{n=1}^{\infty} \left(\frac{|x_n - a_n|}{n} \right)^2$$

$$= \sum_{n=1}^{k} \left(\frac{|x_n - a_n|}{n} \right)^2 + \sum_{n=k+1}^{\infty} \left(\frac{|x_n - a_n|}{n} \right)^2$$

$$< \sum_{n=1}^{k} \left(\frac{2^n}{n} \right)^2 \delta^2 + \frac{\epsilon^2}{2}$$

$$= \frac{\epsilon^2}{2} + \frac{\epsilon^2}{2}$$

$$= \epsilon^2.$$

This shows that f is continuous. □

Corollary 6.3.5. *Let* $\mathbf{H} = \{(x_n) \in \ell_2 \mid 0 \leq x_n \leq \frac{1}{n}\}$. *Then* $\mathbf{H}$ *as a metric subspace of* ℓ_2 *is compact.*

Remark 6.3.6. In view of Proposition 6.3.4, we will also call $\mathbf{H}$ the Hilbert cube.

Proposition 6.3.7. *Every compact metric space is separable.*

Proof. Let X be a compact metric space. Note that for each $n \in \mathbb{N}$, the family

$$\mathcal{U}_n = \left\{ B\left(x, \frac{1}{n} \right) \mid x \in X \right\}$$

is an open cover of X. Since X is compact, there is a finite set A_n in X such that

$$X = \bigcup_{x \in A_n} B\left(x, \frac{1}{n} \right).$$

Let

$$B = \left\{ B\left(x, \frac{1}{n}\right) \,\middle|\, n \in \mathbb{N}, x \in A_n \right\}.$$

Note that B is countable. Let U be an open set, and let $x \in U$. Since U is open, we can find $k \in \mathbb{N}$ such that $B(x, \frac{1}{k}) \subseteq U$. By the property of subcover there is $y \in A_n$ such that $x \in B(y, \frac{1}{n})$. This implies that $B(y, \frac{1}{n}) \in B$ and

$$x \in B\left(y, \frac{1}{n}\right) \subseteq U.$$

Since B is countable, we can renumber B as follows:

$$B = \{B_n \mid n \in \mathbb{N}\}.$$

Choose an element $x_n \in B_n$ and consider the set

$$D = \{x_n \mid n \in \mathbb{N}\}.$$

We can check that D is dense in X. □

Theorem 6.3.8. *A separable metric space can be homeomorphically embedded in the Hilbert cube.*

Proof. Let (X, d) be a separable metric space. Without loss of generality, we can assume that $d(x, y) \leq 1$ for all $x, y \in X$; otherwise, we can define a metric on X that is topologically equivalent to the original metric and bounded by 1. Let

$$A = \{x_n \mid n \in \mathbb{N}\}$$

be a dense set in X. Define the map $f : X \to \mathbf{H}$ by

$$f(x) = \left(\frac{d(x, x_n)}{n}\right).$$

Let $x, y \in X$. Then for $k \in \mathbb{N}$, we have

$$\left| \frac{d(x, x_k)}{k} - \frac{d(y, x_k)}{k} \right| \leq \left(\sum_{n=1}^{\infty} \frac{|d(x, x_n) - d(y, x_n)|^2}{n^2} \right)^{\frac{1}{2}}$$

$$= \|f(x) - f(y)\|_2$$

$$\leq \left(\sum_{n=1}^{\infty} \left(\frac{d(x, y)}{n} \right)^2 \right)^{\frac{1}{2}}, \qquad (6.2)$$

where the last inequality holds because of the inequality

$$\left| d(x,z) - d(y,z) \right| \le d(x,y).$$

We first claim that f is injective. Let $x, y \in X$ be such that $x \ne y$. Since A is dense, there is $x_i \in A$ such that

$$d(x, x_i) < \frac{d(x,y)}{2}.$$

Now if $d(x, x_i) = d(y, x_i)$, then

$$\begin{aligned} d(x,y) &\le d(x, x_i) + d(x_i, y) \\ &= 2d(x, x_i) \\ &< d(x,y). \end{aligned}$$

This is a contradiction. Therefore $d(x, x_i) \ne d(y, x_i)$. Now, by inequality (6.2) we have

$$\|f(x) - f(y)\|_2 \ge \left| \frac{d(x, x_i) - d(y, x_i)}{i} \right| > 0.$$

Therefore $f(x) \ne f(y)$. This shows that f is injective.

Now we claim that f is continuous. Let $\epsilon > 0$ and $a \in X$. Choose $N \in \mathbb{N}$ such that

$$\sum_{n=N+1}^{\infty} \frac{1}{n^2} < \frac{\epsilon^2}{2}.$$

Let $\delta = \frac{\epsilon}{\sqrt{2N}}$. Let $x \in X$ be such that $d(x, a) < \delta$. Then by inequality (6.2) we have

$$\begin{aligned} \|f(x) - f(a)\|_2^2 &= \sum_{n=1}^{\infty} \left(\frac{|d(x, x_n) - d(a, x_n)|}{n} \right)^2 \\ &\le \sum_{n=1}^{\infty} \left(\frac{d(x,a)}{n} \right)^2 \\ &= \sum_{n=1}^{N} \left(\frac{d(x,a)}{n} \right)^2 + \sum_{n=N+1}^{\infty} \left(\frac{d(x,a)}{n} \right)^2 \\ &< \sum_{n=1}^{N} d(x,a)^2 + \frac{\epsilon^2}{2} \\ &< \sum_{n=1}^{N} \frac{\epsilon^2}{2N} + \frac{\epsilon^2}{2} \\ &= \epsilon^2. \end{aligned}$$

Therefore f is continuous at $a \in X$. Since $a \in X$ is arbitrary, f is continuous.

Finally, we claim that f is an open map from X to $f(X)$. Let $\epsilon > 0$ and $a \in X$. To prove that f is open, it is sufficient to prove that the image of an open ball in X is open in $f(X)$.

Since A is dense, there is $x_k \in A$ such that $d(a, x_k) < \frac{\epsilon}{3}$. Let $\delta = \frac{\epsilon}{3k}$. Let $f(x) \in f(X)$ be such that $\|f(x) - f(a)\|_2 < \delta$. By inequality (6.2) we have

$$\left| d(a, x_k) - d(x, x_k) \right| \le k \|f(x) - f(a)\|_2 < \frac{\epsilon}{3}.$$

This implies that

$$d(x, x_k) < d(a, x_k) + \frac{\epsilon}{3} < \frac{2\epsilon}{3}.$$

Then

$$d(x, a) \le d(x, x_k) + d(x_k, a) < \frac{2\epsilon}{3} + \frac{\epsilon}{3} = \epsilon.$$

Therefore $x \in B(a, \epsilon)$. Then $f(x) \in f(B(a, \epsilon))$. This implies that

$$B(f(a), \delta) \cap f(X) \subseteq f(B(a, \epsilon)).$$

Hence f is an open map from X to $f(X)$. Thus f is a homeomorphism from X to $f(X)$. $\square$

Corollary 6.3.9. *A compact metric space can be homeomorphically embedded in the Hilbert cube as a closed subspace.*

Remark 6.3.10. Since the cardinality of $[0,1]^{\mathbb{N}}$ is equal to that of $[0,1]$, we can observe that if X is a metric space with cardinality strictly greater than that of $[0,1]$, then X cannot be a compact metric space.

6.4 Cantor set

For a given closed interval $A = [a, b]$, let $\acute{A}$ denote the set after deleting the open middle third interval of A, that is,

$$\acute{A} = \left[a, a + \frac{1}{3}(b-a) \right] \cup \left[a + \frac{2}{3}(b-a), b \right].$$

Also, for a given union

$$B = \bigcup_{i=1}^{n} A_i$$

of finite disjoint closed intervals $A_i = [a_i, b_i]$, let $\acute{B} = \bigcup_{i=1}^{n} \acute{A}_i$.

Consider the interval $[0, 1]$. We will inductively apply the process of deleting the open middle third interval. Let

$$C_0 = [0, 1].$$

Then define $C_1 = \acute{C}_0$, that is,

$$C_1 = \left[0, \frac{1}{3}\right] \cup \left[\frac{2}{3}, 1\right].$$

Now define $C_2 = \acute{C}_1$, that is,

$$C_2 = \left[0, \frac{1}{9}\right] \cup \left[\frac{2}{9}, \frac{1}{3}\right] \cup \left[\frac{2}{3}, \frac{7}{9}\right] \cup \left[\frac{8}{9}, 1\right].$$

When C_n is defined, define $C_{n+1} = \acute{C}_n$. Note that each C_n is the union of 2^n closed intervals. Therefore each C_n is closed. Now define

$$C = \bigcap_{n=1}^{\infty} C_n.$$

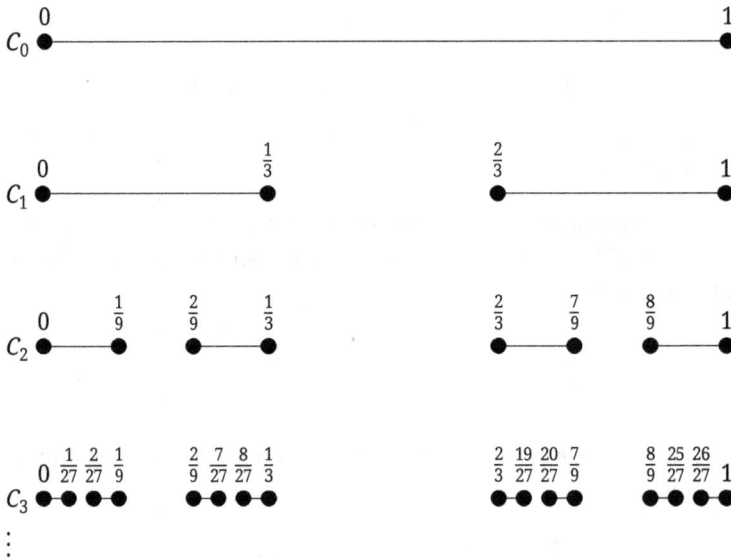

Construction of the Cantor set

Consider the set C with the metric induced by the usual metric of $\mathbb{R}$. The metric space C is called the Cantor ternary set or simply the Cantor set. Since an arbitrary intersection of closed sets is closed, C is closed. Also, C is contained in $[0, 1]$, so C is compact. We will say that $x \in C$ is an end point of C if it is an end point of one of the open intervals removed from C_n to obtain C_{n+1}.

Proposition 6.4.1. *No singleton is open in the Cantor set.*

Proof. On the contrary, suppose that there is a point $x \in C$ such that $\{x\}$ is open in the Cantor set C. Then there is $r > 0$ such that

$$(x - r, x + r) \cap C = \{x\}. \tag{6.3}$$

Choose $k \in \mathbb{N}$ such that $\frac{1}{3^k} < r$. Note that $x \in C_k$. Let $[a, b]$ be an interval of C_k such that $x \in [a, b]$. Then

$$[a, b] \subseteq (x - r, x + r).$$

Note that one of the end points y of $[a, b]$ is different from x. Since $y \in C$, we get a contradiction to equation (6.3). $\qquad \square$

By Proposition 6.4.1 we observe that the Cantor set is different from the discrete metric space.

Proposition 6.4.2. *For the Cantor set C, $D(C) = C$.*

Proof. Since C is closed, $D(C) \subseteq C$. Conversely, let $x \in C$. Consider an open ball $B(x, r)$. Choose $n \in \mathbb{N}$ such that $\frac{1}{3^n} < r$. Then $B(x, r)$ intersects C_n, and therefore it intersects C in a point other than x. Hence $x \in D(C)$. This shows that $D(C) = C$. $\qquad \square$

By definition the Cantor set C is a subset of $[0, 1]$. Let us try to find out which points of $[0, 1]$ are in the Cantor set. We have observed that every element in $[0, 1]$ can be represented in the base $b > 1$. Let us take $b = 3$. Let $x \in [0, 1]$. Then

$$x = (0.x_1 x_2 \cdots x_n \cdots)_3.$$

If $x = 0$, then $x_n = 0$ for all $n \in \mathbb{N}$. Let $x > 0$. Suppose we have the following situation in the base 3 representation:

$$x_m = 1 \quad \text{for some } m, \quad \text{and} \quad x_n = 0 \quad \text{for all } n > m. \tag{6.4}$$

Then we can choose $x_n \in \{0, 1, 2\}$ as follows to avoid the above situation.
Choose the unique $x_1 \in \{0, 1, 2\}$ such that

$$\frac{x_1}{3} < x \le \frac{x_1 + 1}{3}.$$

Suppose we have obtained x_{n-1}. Choose the unique $x_n \in \{0, 1, 2\}$ such that

$$\frac{x_n}{3^n} < x - \sum_{i=1}^{n-1} \frac{x_i}{3^i} \le \frac{x_n + 1}{3^n}.$$

On the contrary, suppose (6.4) holds. Then

$$x - \sum_{i=1}^{m-1} \frac{x_i}{3^i} = \frac{1}{3^m}.$$

Now by the construction we have

$$\frac{1}{3^m} < x - \sum_{i=1}^{m-1} \frac{x_i}{3^i} = \frac{1}{3^m},$$

a contradiction.

Now suppose we have the following situation in the base 3 representation:

$$x_m = 1 \quad \text{for some } m, \quad \text{and} \quad x_n = 2 \quad \text{for all } n > m. \tag{6.5}$$

Then we can redefine x_n for $n \geq m$ as follows. Take $x_m = 2$ and $x_n = 0$ for all $n > m$. We claim that for all $n \in \mathbb{N}$, $x \in C_n$ if and only if all $x_1, x_2, \ldots, x_n$ are different from 1. The proof of the claim is by induction on n.

Let $x \in [0, \frac{1}{3}] \cup [\frac{2}{3}, 1]$. Then either $x \in [0, \frac{1}{3}]$ or $x \in [\frac{2}{3}, 1]$. Then

$$x = (0.0x_2x_3 \cdots)_3 \quad \text{or} \quad x = (0.2x_2x_3 \cdots)_3.$$

This shows that the statement is true for $n = 1$. Since the middle one third of the subintervals is removed in each step, we can similarly observe that the statement holds inductively. Thus we have shown that x is in the Cantor set if and only if each x_n in the base 3 representation is either 0 or 2. This defines the bijective map $\kappa : C \to \{0, 2\}^{\mathbb{N}}$ by

$$\kappa(x) = (0.x_1x_2 \cdots)_3,$$

where $x_n \in \{0, 2\}$. Consider $\{0, 2\}$ with the discrete metric and equip $\{0, 2\}^{\mathbb{N}}$ with the metric defined by

$$d(x, y) = \sum_{i=1}^{\infty} \frac{d_i(x_i, y_i)}{2^i}.$$

We claim that κ is a homeomorphism. To prove this, we will prove that its inverse map is a homeomorphism. Let κ' be the inverse of κ. Then

$$\kappa'((0.x_1 \cdots x_n \cdots)_3) = \sum_{n=1}^{\infty} \frac{x_n}{3^n}.$$

We first show that κ' is continuous. Let $\epsilon > 0$. Choose $n \in \mathbb{N}$ such that $\frac{1}{3^n} < \epsilon$. Let

$$a = (0.a_1a_2 \cdots)_3 \in \{0, 2\}^{\mathbb{N}}$$

be such that

$$a \in \{a_1\} \times \{a_2\} \times \cdots \times \{a_n\} \times \{0, 2\} \times \{0, 2\} \times \cdots.$$

Let

$$U_n = \{a_1\} \times \{a_2\} \times \cdots \times \{a_n\} \times \{0, 2\} \times \{0, 2\} \times \cdots.$$

Note that U_n is an open set in $\{0, 2\}^{\mathbb{N}}$. By Proposition 4.1.6, if U is an open set in $\{0, 2\}^{\mathbb{N}}$ containing a, then

$$a \in U_n \subseteq U.$$

Now for $x \in U_n$, we have

$$
\begin{aligned}
\left| \kappa'(x) - \kappa'(a) \right| &= \left| \sum_{i=n+1}^{\infty} \frac{x_i - a_i}{3^i} \right| \\
&\leq \sum_{i=n+1}^{\infty} \frac{2}{3^i} \\
&= \frac{1}{3^n} \\
&< \epsilon.
\end{aligned}
$$

This shows that κ' is continuous at a. Since $a \in \{0, 2\}^{\mathbb{N}}$ is arbitrary, κ' is continuous.

Note that since $\{0, 2\}$ is closed in $[0, 1]$, $\{0, 2\}^{\mathbb{N}}$ is closed in the Hilbert cube $[0, 1]^{\mathbb{N}}$. Since the Hilbert cube is compact, $\{0, 2\}^{\mathbb{N}}$ is compact. Since κ' is a bijective continuous map from a compact set, κ' is a homeomorphism. This shows that the Cantor set is homeomorphic to the countable product of the discrete metric spaces consisting of two elements. By Proposition 6.4.1, $\{0, 2\}^{\mathbb{N}}$ is not a discrete metric space. Let us observe a few more properties of the Cantor set.

Proposition 6.4.3. *The Cantor set is homeomorphic to the countable product of the Cantor sets.*

Proof. From the fact that the cardinality of $\mathbb{R}^{\mathbb{N}}$ is equal to that of $\mathbb{R}$ we can observe that the countable product of metric spaces each of which is a countable product of the sets $\{0, 2\}$ is a countable product of the sets $\{0, 2\}$. This shows that the Cantor set is homeomorphic to the countable product of the Cantor sets. $\square$

Proposition 6.4.4. *The closed interval $[0, 1]$ is the continuous image of the Cantor set.*

Proof. Let $x = (0.x_1 \cdots x_n \cdots)_3 \in C$. Then each $x_n \in \{0, 2\}$. Define

$$y = (0.y_1 \cdots y_n \cdots)_2, \quad \text{where } y_n = \frac{x_n}{2}.$$

Define the map $f : C \to [0, 1]$ by $f(x) = y$. Clearly, the map f is surjective as when we double the binary digits of elements of $[0, 1]$, we get a member of the Cantor set C represented in the base 3.

Now we prove that f is continuous. Let $x, y \in C$. Suppose that $|x - y| < \frac{1}{3^n}$. Then x and y lie in the same nth step subinterval. This implies that the even representations of x and y in the base 3 coincide through the nth digit. Therefore the binary representations of $f(x)$ and $f(y)$ coincide through the nth digit. Hence $|f(x) - f(y)| < \frac{1}{2^n}$.

Let $\epsilon > 0$. Choose $n \in \mathbb{N}$ such that $\frac{1}{2^n} < \epsilon$. Also, choose $\delta > 0$ such that $\delta < \frac{1}{3^n}$. Then

$$|x - y| < \delta \Rightarrow |f(x) - f(y)| < \epsilon.$$

This shows that f is continuous. □

Remark 6.4.5. Let us consider the endpoints of the first step, $\frac{1}{3}$ and $\frac{2}{3}$. Then

$$\frac{1}{3} = (0.022\cdots)_3 \quad \text{and} \quad \frac{2}{3} = (0.200\cdots)_3.$$

These elements are mapped to the same element $\frac{1}{2}$ as

$$\frac{1}{2} = (0.011\cdots)_2 = (0.100\cdots)_2.$$

We have the same situation for endpoints of each step.

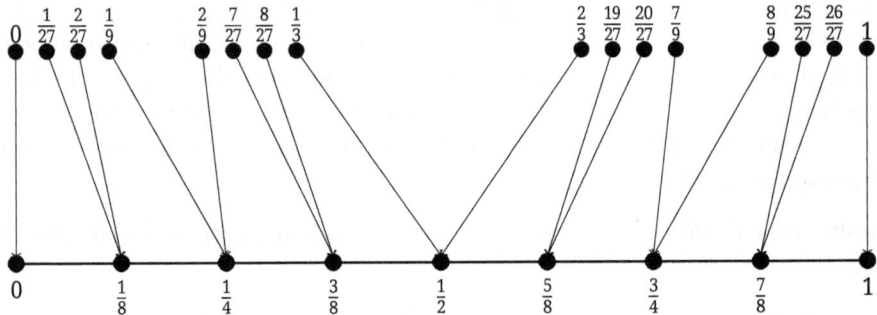

Continuous surjection from C to $[0, 1]$

Corollary 6.4.6. *The Hilbert cube $[0, 1]^{\mathbb{N}}$ is the continuous image of the Cantor set.*

Proof. Consider the countable product $C^{\mathbb{N}}$ of the Cantor sets C. Let $f : C \to [0, 1]$ be a surjective continuous map. Define the map $\phi : C^{\mathbb{N}} \to [0, 1]^{\mathbb{N}}$ by

$$\phi(x_1, x_2, \ldots) = (f(x_1), f(x_2), \ldots).$$

We can check that ϕ is a surjective continuous map. Since $C^{\mathbb{N}}$ is homeomorphic to C, the Hilbert cube $[0, 1]^{\mathbb{N}}$ is the continuous image of the Cantor set. □

Let C' denote the set obtained as the intersection of the sets constructed out of $[0, 1]$ after removing the middle two-third in each step as we did to obtain the Cantor set C. By a similar discussion we can show that C' consists of all the points

$$x = (0.x_1 x_2 \cdots x_n \cdots)_6$$

of $[0,1]$ such that each x_n is either 0 or 5. We can easily observe that C' is homeomorphic to the Cantor set C.

Theorem 6.4.7. *Every closed subset of the Cantor set is the continuous image of the Cantor set.*

Proof. Let F be a closed subset of the set C' discussed above. Note that if $x, y \in C'$, then the midpoint $\frac{x+y}{2} \notin C'$. Let $x \in C'$. Then there is a unique point $y_x \in F$ that is closest to x. This defines the map $f : C' \to F$ by $f(x) = y_x$. We can check that f is surjective and continuous. Since the Cantor set C is homeomorphic to C', F is the continuous image of the Cantor set. $\square$

Corollary 6.4.8. *Every compact metric space is the continuous image of the Cantor set.*

Proof. Let X be a compact metric space. Then X is embedded in the Hilbert cube $[0,1]^{\mathbb{N}}$. Without loss of generality, we can suppose that X is a closed set in $[0,1]^{\mathbb{N}}$. Let $\phi : C^{\mathbb{N}} \to [0,1]^{\mathbb{N}}$ be a surjective continuous map. Then $\phi^{-1}(X)$ is a closed set in the Cantor set C. By Theorem 6.4.7, $\phi^{-1}(X)$ is the continuous image of the Cantor set C. Since X is the continuous image of $\phi^{-1}(X)$, X is the continuous image of the Cantor set. $\square$

Exercises

6.1. First of all, complete whatever is left for you as exercises.

6.2. Suppose that there is a nonzero linear functional f on a vector space V that is continuous with respect to all the norms on V. Can you conclude that V is finite dimensional?

6.3. Show that the finite union of compact sets in a metric space is compact.

6.4. Show that the arbitrary intersection of nonempty compact sets in a metric space is compact.

6.5. Show that if X is a compact metric space such that each point of X is a limit point, then X is uncountable.

6.6. Check whether the closed ball $D(0,1)$ is sequentially compact in ℓ_∞ or not.

6.7. Let (X, d) be a compact metric space. Let $f : X \to X$ be a map such that $d(f(x), f(y)) < d(x, y)$ for all $x, y \in X$. Show that there is a unique fixed point of f.

6.8. Consider the normed space $(C[0,1], \|\cdot\|_\infty)$. Show that the set

$$A = \{f \in C[0,1] \mid f([0,1]) \subseteq [0,1]\}$$

is closed and bounded but not compact in $(C[0,1], \|\cdot\|_\infty)$.

6.9. Let A and B be closed and compact sets in a metric space X such that $A \cap B = \emptyset$. Show that $d(A, B) > 0$.

6.10. Let A be compact set in a metric space X. Show that there are $x, y \in X$ such that $\operatorname{diam}(A) = d(x, y)$.

6.11. Show that a metric space X is compact if and only if X is complete with respect to all the topologically equivalent metrics on it.

6.12. What are compact sets in $L^1[0, 1]$?

6.13. (Arzelà–Ascoli theorem) A family $\mathcal{A}$ of maps from a metric space X to a metric space Y is called equicontinuous at $a \in X$ if for each $\epsilon > 0$, there is $\delta > 0$ such that for each $f \in \mathcal{A}$,

$$d(x, a) < \delta \Rightarrow d(f(x), f(a)) < \epsilon.$$

The family $\mathcal{A}$ is called equicontinuous if it equicontinuous at each point of X.

Let X be a compact metric space. Show that $\mathcal{A}$ is compact in the normed space $(C_{\mathbb{F}}(X), \|\cdot\|_\infty)$ of all continuous maps from X to $\mathbb{F}$ if and only if $\mathcal{A}$ is closed, bounded, and equicontinuous.

6.14. Show that a metric space X is totally bounded if and only if its completion $\hat{X}$ is compact.

6.15. Let X be a metric space. Then a family $\mathcal{A}$ of nonempty sets in X is said to have the finite intersection property if every finite subfamily of $\mathcal{A}$ has a nonempty intersection.

Show that a metric space X is compact if and only if every family of closed sets with finite intersection property has a nonempty intersection.

7 Connected metric spaces

In this chapter, we study connected metric spaces and their properties.

7.1 Connectedness

Consider the real line $\mathbb{R}$. Suppose we wish to express the real line as a union of two nonempty disjoint open sets U and V, that is,

$$\mathbb{R} = U \cup V \quad \text{and} \quad U \cap V = \emptyset.$$

Choose $a \in U$ and $b \in V$ and suppose that $a < b$. Consider the set

$$A = \{x \in \mathbb{R} \mid [a, x) \subseteq U\}.$$

Note that $a \in A$ and b is an upper bound of A. By the order completeness of $\mathbb{R}$, the supremum of A exists in $\mathbb{R}$. Let $l = \sup A$. We can check that l is a limit point of U. Since U is the complement of V, U is closed in $\mathbb{R}$. This implies that $l \in U$. Also, since U is open, there is $r > 0$ such that

$$(l - r, l + r) \subseteq U.$$

This implies that $[l - \frac{r}{2}, l + \frac{r}{2}) \subseteq U$, which shows that $[a, l + \frac{r}{2}) \subseteq U$. This is a contradiction. Therefore the real line cannot be represented as a union of two nonempty disjoint open sets.

Definition 7.1.1. A metric space X is called connected if it cannot be expressed as a union of two nonempty disjoint open sets.

A metric space X is called disconnected if it is not connected. Suppose a metric space X is disconnected. Then there are nonempty disjoint open sets U and V such that

$$X = U \cup V.$$

The pair (U, V) is called a disconnection of X.

Example 7.1.2. The real line is connected.

Example 7.1.3. The discrete metric space X containing more than one point is disconnected. For any two distinct points x and y, there is a disconnection, one containing x and the other containing y.

Proposition 7.1.4. *A metric space X is disconnected if and only if there exist two nonempty disjoint sets A and B in X such that $X = A \cup B$, $(\mathrm{Cl}\,A) \cap B = \emptyset$, and $A \cap (\mathrm{Cl}\,B) = \emptyset$.*

https://doi.org/10.1515/9783111636085-007

Proof. Suppose that X is disconnected. Then there are nonempty disjoint open sets U and V such that $X = U \cup V$. Since U and V are closed, $\operatorname{Cl} U = U$ and $\operatorname{Cl} V = V$. Therefore $(\operatorname{Cl} U) \cap V = \emptyset$, and $U \cap (\operatorname{Cl} V) = \emptyset$.

Conversely, suppose that there exist two nonempty disjoint sets A and B in X such that $X = A \cup B$, $(\operatorname{Cl} A) \cap B = \emptyset$, and $A \cap (\operatorname{Cl} B) = \emptyset$. We claim that (A, B) is a disconnection of X. Note that

$$B \subseteq X \setminus \operatorname{Cl} A \subseteq X \setminus A = B.$$

This implies that $\operatorname{Cl} A = A$. Similarly, $\operatorname{Cl} B = B$. This shows that (A, B) is a disconnection of X. □

Proposition 7.1.5. *A metric space X is connected if and only if every continuous map $f : X \to \{0, 1\}$ is constant.*

Proof. Suppose that X is connected. If there is a continuous map $f : X \to \{0, 1\}$ that is not constant, then the pair $(f^{-1}\{0\}, f^{-1}\{1\})$ forms a disconnection. This is a contradiction.

For the converse, suppose that X is disconnected. Then there are nonempty disjoint open sets U and V such that $X = U \cup V$. Define the map $f : X \to \{0, 1\}$ by $f(U) = \{0\}$ and $f(V) = \{1\}$. Note that f is continuous and nonconstant. □

Proposition 7.1.6. *Let A be a connected set in a metric space X. Let $A \subseteq B \subseteq \operatorname{Cl} A$. Then B is connected.*

Proof. Let $f : B \to \{0, 1\}$ be a continuous map. Since A is connected, f is constant on A. Suppose that $f(A) = \{1\}$. Let $b \in B$. Note that $\{f(b)\}$ is open in $\{0, 1\}$. Since f is continuous at b, there is an open set U containing b such that $f(U) \subseteq \{f(b)\}$. Note that $B \subseteq \operatorname{Cl} A = A \cup D(A)$. If $b \in A$, then $f(b) = 1$. If $b \in D(A)$, then $A \cap (U \setminus \{b\}) \neq \emptyset$. Let $a \in A \cap (U \setminus \{b\})$. Then $f(a) \in \{f(b)\}$. This implies that $f(b) = f(a) = 1$. Therefore, f is constant on B. Hence B is connected. □

Proposition 7.1.7. *A metric space X is connected if and only if the empty set $\emptyset$ and X are the only clopen sets of X.*

Proof. Suppose that X is connected. On the contrary, suppose that U is a clopen set different from $\emptyset$ and X. Let $V = X \setminus U$. Then V is a nonempty open set disjoint from U. Since $U \cup V = X$, X is disconnected. This is a contradiction.

Conversely, suppose that X has no nonempty proper clopen set. On the contrary, suppose that X is disconnected. Then there are nonempty disjoint open sets U and V such that

$$X = U \cup V \quad \text{and} \quad U \cap V = \emptyset.$$

This shows that U is a clopen set in X, a contradiction. □

Proposition 7.1.8. *The continuous image of a connected metric space is connected.*

Proof. Suppose that f is a surjective continuous map from a connected metric space X to a metric space Y. On the contrary, suppose that Y is disconnected. Then there are nonempty disjoint open sets U and V such that

$$Y = U \cup V \quad \text{and} \quad U \cap V = \emptyset.$$

We can easily observe that the pair $(f^{-1}(U), f^{-1}(V))$ is a disconnection of X, a contradiction. $\quad\square$

Corollary 7.1.9. *Let X be a connected metric space. If Y is a metric space homeomorphic to X, then Y is connected.*

A set A in a metric space X is called connected if it is connected as a metric subspace of X. Note that each singleton set in a metric space is a connected set.

Example 7.1.10. Define the map $f : \mathbb{R} \to \mathbb{R}^2$ by $f(x) = (\cos x, \sin x)$. We can check that f is continuous. Note that the image of f is the unit circle $\mathbb{S}^1$. By Proposition 7.1.8, $\mathbb{S}^1$ is a connected set in $\mathbb{R}^2$.

Now we will classify the connected sets in the real line.

A set I in the real line $\mathbb{R}$ is called an interval if for all $x, y \in I$ and for every $z \in \mathbb{R}$ with $x < z < y$, we have $z \in I$. We leave as an exercise to classify the intervals in $\mathbb{R}$.

Proposition 7.1.11. *A set in the real line is connected if and only if it is an interval.*

Proof. Suppose that I is a connected set in the real line $\mathbb{R}$. Suppose that I is not an interval. Then there are points $a, b \in I$ with $a < b$ such for some point c with $a < c < b, c \notin I$. Note that

$$I = (I \cap (-\infty, c)) \cup (I \cap (c, \infty)).$$

This shows that I is disconnected, a contradiction.

Conversely, suppose that I is an interval in the real line $\mathbb{R}$. On the contrary, suppose that I is disconnected. Suppose that (A, B) is a disconnection of I. Choose $a \in A$ and $b \in B$ such that $a < b$. Consider the set

$$Y = \{x \in \mathbb{R} \mid [a, x) \subseteq A\}.$$

Note that $Y \neq \emptyset$ and Y is bounded above by b. Let $l = \sup Y$. Since $a \leq l \leq b$ and I is an interval, $l \in I$. Note that l is a limit point of A. Then

$$l \in \mathrm{Cl}_X A \cap I = \mathrm{Cl}_I A.$$

Since A is closed in I, $\mathrm{Cl}_I A = A$. This shows that $l \in A$. Therefore $l < b$. Since A is open in I, there is $\epsilon > 0$ such that

$$(l - \epsilon, l + \epsilon) \cap I \subseteq A.$$

Since

$$(l - \epsilon, l + \epsilon) \cap [a, b] \subseteq (l - \epsilon, l + \epsilon) \cap I \subseteq A \quad \text{and} \quad l < b,$$

we get a contradiction as l is the supremum of Y. □

Proposition 7.1.12 (Intermediate value theorem). *Let X be a connected metric space, and let $f : X \to \mathbb{R}$ be a continuous map. Let $a, b \in f(X)$ with $a < b$. If $c \in \mathbb{R}$ is such that $a < c < b$, then there exists $x \in X$ such that $f(x) = c$.*

Proof. Note that $f(X)$ is connected in $\mathbb{R}$. By Proposition 7.1.11, $f(X)$ is an interval in $\mathbb{R}$. Therefore $c \in f(X)$. Hence there exists $x \in X$ such that $f(x) = c$. □

Proposition 7.1.13. *Let X be a connected metric space containing more than one element. Then X is uncountable.*

Proof. Let $a, b \in X$ be such that $a \neq b$. Define the map $f : X \to \mathbb{R}$ by $f(x) = d(a, x)$. Then f is continuous. Therefore $f(X)$ is connected in $\mathbb{R}$. By Proposition 7.1.11, $f(X)$ is an interval in $\mathbb{R}$. Note that $f(a) = 0$ and $f(b) = d(a, b) > 0$. Since $f(X)$ is interval, $[0, d(a, b)] \subseteq f(X)$. Therefore $f(X)$ is uncountable. □

Suppose that A is a connected set in a metric space X such that $A \subseteq U \cup V$, where U and V are open sets with $U \cap V \cap A = \emptyset$. Then $A = (U \cap A) \cup (V \cap A)$. Since A is connected, either $U \cap A = A$, or $V \cap A = A$. This implies that $A \subseteq U$ or $A \subseteq V$.

Proposition 7.1.14. *A function $f : [a, b] \to \mathbb{R}$ is monotonic if and only if $f^{-1}(A)$ is connected for every connected set A in $\mathbb{R}$.*

Proof. Suppose f is increasing. Let A be a connected set in $\mathbb{R}$. On the contrary, suppose that $f^{-1}(A)$ is disconnected. Then there are points $x, y \in f^{-1}(A)$ and $z \notin f^{-1}(A)$ such that $x < z < y$. Then $f(x) \leq f(z) \leq f(y)$ and $f(z) \notin A$. This is a contradiction. We can similarly discuss the decreasing case.

Conversely, suppose that $f^{-1}(A)$ is connected for every connected set A in $\mathbb{R}$. On the contrary, suppose that f is not monotonic. Then f is neither increasing not decreasing. Since f is not increasing, there are points $x, y \in [a, b]$ such that $x < y$ and $f(x) > f(y)$. Since f is not decreasing, there is a point $z \in [a, b]$ such that one of the following holds:
(i) $z > y$ and $f(z) > f(y)$;
(ii) $y > z > x$ and $f(z) > f(x)$;
(iii) $x > z$ and $f(x) > f(z)$.

If (i) or (ii) holds, then $f^{-1}(f(y), \infty)$ is disconnected. If (iii) holds, then $f^{-1}(-\infty, f(x))$ is disconnected. In all the cases, we have a contradiction. □

Proposition 7.1.15. *Let $\{A_\alpha \mid \alpha \in \mathcal{A}\}$ be a nonempty family of connected sets in a metric space X such that their intersection is nonempty. Then $\bigcup_{\alpha \in \mathcal{A}} A_\alpha$ is connected.*

Proof. Let $A = \bigcup_{\alpha \in \mathcal{A}} A_\alpha$. On the contrary, suppose that A is disconnected. Then there are open sets U and V of X such that $A = (U \cap A) \cup (V \cap A)$, where $U \cap A \neq \emptyset$, $V \cap A \neq \emptyset$, and $U \cap V \cap A = \emptyset$. Since for each $\alpha \in \mathcal{A}$, A_α is connected and $A_\alpha \subseteq A$, we have either $A_\alpha \subseteq U$ or $A_\alpha \subseteq V$ for all $\alpha \in \mathcal{A}$.

Let $x \in \bigcap_{\alpha \in \mathcal{A}} A_\alpha$. Since $A \subseteq U \cup V$, either $x \in U$, or $x \in V$. Suppose that $x \in U$. Then by the above observation we have $A_\alpha \subseteq U$ for all $\alpha \in \mathcal{A}$. Since $U \cap V \cap A = \emptyset$, we have $V \cap A = \emptyset$, a contradiction. □

Proposition 7.1.16. *Let X and Y be connected metric spaces. Then $X \times Y$ is connected.*

Proof. Let $x \in X$. Define the map $f : Y \to \{x\} \times Y$ by $f(y) = (x, y)$. By Proposition 7.1.8, $\{x\} \times Y$ is connected in $X \times Y$. Similarly, $X \times \{y\}$ is connected in $X \times Y$ for all $y \in Y$.

Let $(a, b) \in X \times Y$. For each $y \in Y$, consider the set

$$A_y = (\{a\} \times Y) \cup (X \times \{y\}).$$

By Proposition 7.1.15, A_y is connected for all $y \in Y$. Since $(a, b) \in A_y$ for each $y \in Y$, by Proposition 7.1.15, $X \times Y = \bigcup_{y \in Y} A_y$ is connected. □

7.2 Components in metric spaces

A maximal connected subset of a metric space X is called a component of X. Let $a \in X$. Let $\mathcal{A}$ be the collection of connected sets in X containing a. Note that $\mathcal{A}$ is nonempty set as $\{a\}$ is a connected set. Let C be the union of all members of $\mathcal{A}$. By Proposition 7.1.15, C is connected. Note that C is a component of X.

Let A be a connected set in X, and let $a \in A$. Let C be a component of X containing a. Then by Proposition 7.1.15, $A \cup C$ is connected. By the maximality, $A \cup C = C$. This shows that $A \subseteq C$.

Let C_1 and C_2 be distinct components of X. Then $C_1 \cap C_2 = \emptyset$; otherwise, $C_1 \cup C_2$ would be connected, which is a contradiction. Thus we have observed that the collection of components forms a partition of X.

If A is a connected set, then $\text{Cl}\, A$ is a connected set. This shows that if A is a component of X, then $\text{Cl}\, A = A$. Therefore A is closed.

Example 7.2.1. Each singleton is a component in the discrete metric space.

Example 7.2.2. Consider the set $X = \{0\} \cup \{\frac{1}{n} \mid n \in \mathbb{N}\}$ in the real line. Then each singleton in X is a component of X. Note that $\{0\}$ is not open in X.

Example 7.2.3. Let U be an open set in the real line. Let C be a component of U. Let $a \in C \subseteq U$. Since U is open, there is $r > 0$ such that $B(a, r) \subseteq U$. Since $a \in B(a, r) \cap C$,

$B(a,r) \cup C$ is connected. This implies that $B(a,r) \cup C = C$. Therefore $B(a,r) \subseteq C$. This shows that C is open in $\mathbb{R}$. Since each component of U contains a point of $U \cap \mathbb{Q}$, there are countably many components of U.

Proposition 7.2.4. *If A is a connected clopen set in a metric space X, then A is a component of X.*

Proof. Let C be a component of X such that $A \subseteq C$. If $A \neq C$, then $A \cap C$ is a nonempty proper clopen set in A. This shows that A is disconnected. This is a contradiction. □

Definition 7.2.5. A metric space X is said to be totally disconnected if for each distinct points x and y of X, there is a disconnection (A, B) of X such that $x \in A$ and $y \in B$.

A set in a metric space X is totally disconnected if it is totally disconnected as a metric subspace.

Example 7.2.6. The discrete metric space is totally disconnected.

Example 7.2.7. Let x and y be two rational numbers such that $a < b$. Let c be an irrational number such that $a < c < b$. Note that the sets

$$A = \{x \in \mathbb{Q} \mid x < c\} \quad \text{and} \quad B = \{x \in \mathbb{Q} \mid x > c\}$$

form a disconnection of $\mathbb{Q}$. Since $a \in A$ and $b \in B$, the set of rational numbers $\mathbb{Q}$ is totally disconnected.

Example 7.2.8. The Cantor set C is totally disconnected. Let x and y be two distinct points of C. Choose $n \in \mathbb{N}$ such that $\frac{1}{3^n} < |x-y|$. Then x and y belong to different subintervals of C_n. Let I be the closed subinterval of C_n such that $x \in I$. Then $(C \cap I, C \backslash I)$ is a disconnection of C such that $x \in C \cap I$ and $y \in C \backslash I$.

Proposition 7.2.9. *The only components of a totally disconnected metric space are singletons.*

Proof. Let C be a component of a totally disconnected metric space X such that C contains distinct points x and y. Then there exists a disconnection (A, B) of X such that $x \in A$ and $y \in B$. Note that $(C \cap A, C \cap B)$ is a disconnection of C. This is a contradiction. □

Proposition 7.2.10. *Let X be a metric space. Suppose that $\mathcal{B}$ is a collection of clopen sets of X such that for any open set U of X with $x \in U$, there is $B \in \mathcal{B}$ such that $x \in B \subseteq U$. Then X is totally disconnected.*

Proof. Let x and y be distinct points of X. Let U be an open set of X such that $x \in U$ and $y \notin U$. Then there is $B \in \mathcal{B}$ such that $x \in B \subseteq U$. Note that $(B, X \backslash B)$ is a disconnection of X such that $x \in B$ and $y \in X \backslash B$. Therefore X is totally disconnected. □

Remark 7.2.11. If X is a compact and totally disconnected metric space, then there is a collection $\mathcal{B}$ of clopen sets of X such that for any open set U of X with $x \in U$, there is $B \in \mathcal{B}$ such that $x \in B \subseteq U$. We leave the proof as an exercise.

7.3 Path connectedness

Let X be a metric space, and let $x, y \in X$. A continuous map $\gamma : [a, b] \to X$ such that $\gamma(a) = x$ and $\gamma(b) = y$ is called a path in X from x to y.

A metric space X is called path connected if for all points x and y of X, there is a path from x to y. A set A in a metric space X is called path connected if it is path connected as a metric subspace.

Example 7.3.1. For $n \geq 2$, $\mathbb{R}^n \setminus \{0\}$ is path connected. Let $x, y \in \mathbb{R}^n \setminus \{0\}$. If the origin is not on the line in $\mathbb{R}^n$ from x to y, then it is a path in $\mathbb{R}^n \setminus \{0\}$ from x to y. If the origin is on the line in $\mathbb{R}^n$ from x to y, then consider a point $z \in \mathbb{R}^n \setminus \{0\}$ that is not on this line. Consider a line from x to z and a line from z to y. This gives a path from x to y.

Example 7.3.2. Let $x, y,$ and z be three points in a metric space X. Let $\gamma_1 : [0, 1] \to X$ be a path from x to y, and let $\gamma_2 : [0, 1] \to X$ be a path from y to z. Define $\gamma : [0, 1] \to X$ by

$$\gamma(t) = \begin{cases} \gamma_1(2t) & \text{if } t \in [0, \tfrac{1}{2}], \\ \gamma_2(2t - 1) & \text{if } t \in [\tfrac{1}{2}, 1]. \end{cases}$$

Then γ is a path in X from x to z.

Remark 7.3.3. By Example 7.3.2 we can observe that a metric space X is path connected if an only if there is a point $a \in X$ that can be joined by a path to any $x \in X$.

Proposition 7.3.4. *A path-connected metric space is connected.*

Proof. Let X be a path-connected metric space. On the contrary, suppose that X is not connected. Let (A, B) be a disconnection of X. Let $x \in A$ and $y \in B$. Since X is path connected, there is a path $\gamma : [a, b] \to X$ such that $\gamma(a) = x$ and $\gamma(b) = y$. Since γ is continuous, $(\gamma^{-1}(A), \gamma^{-1}(B))$ forms a disconnection of $[a, b]$. This shows that $[a, b]$ is disconnected. This is a contradiction. □

Example 7.3.5. The space $\mathbb{R} \setminus \{0\}$ is not path connected as it is not connected.

Proposition 7.3.6. *The continuous image of a path-connected metric space is path connected.*

Proof. Let X and Y be metric spaces such that X is path connected. Let $f : X \to Y$ be a surjective continuous map. Let $y_1, y_2 \in Y$. Then there exist $x_1, x_2 \in X$ such that $f(x_1) = y_1$ and $f(x_2) = y_2$. Since X is path connected, there is a path $\gamma : [a, b] \to X$ such that

$\gamma(a) = x_1$ and $\gamma(b) = x_2$. Then $f \circ \gamma : [a, b] \to X$ is a path from y_1 to y_2. Therefore Y is path connected. □

Example 7.3.7. For $n \geq 1$, consider the sphere $\mathbb{S}^n$. The map $f : \mathbb{R}^{n+1} \setminus \{0\} \to \mathbb{S}^n$ defined by $f(x) = \frac{x}{\|x\|}$ is a surjective continuous map. Since $\mathbb{R}^{n+1} \setminus \{0\}$ is path connected, $\mathbb{S}^n$ is path connected.

A connected metric space need not be path connected as observed in the following example.

Example 7.3.8. Consider the set

$$A = \left\{ (x, y) \in \mathbb{R}^2 \,\middle|\, y = \sin \frac{1}{x}, 0 < x \leq 1 \right\}.$$

We will first show that A is path connected but $\mathrm{Cl}\, A$ is not path connected. Once we obtain that A is path connected, we get that A is connected. This implies that $\mathrm{Cl}\, A$ is connected. Hence $\mathrm{Cl}\, A$ is connected but not path connected (see Figure 7.1).

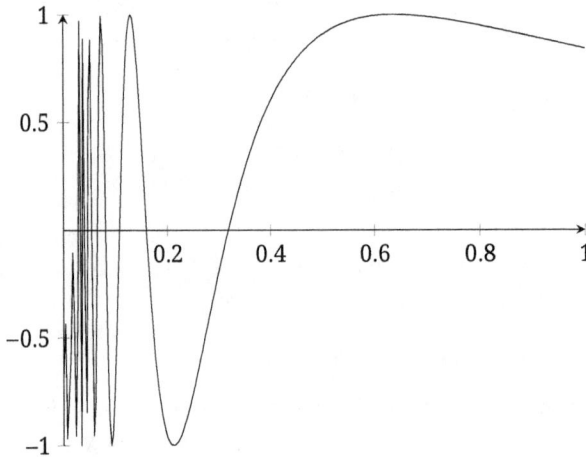

Figure 7.1: Topologist's sine curve.

First, note that A is the continuous image of the map $f : (0, 1] \to \mathbb{R}^2$ defined by

$$f(x) = \left(x, \sin \frac{1}{x} \right).$$

Since $(0, 1]$ is path connected, A is path connected. Observe that

$$\mathrm{Cl}\, A = A \cup \{ (0, y) \mid -1 \leq y \leq 1 \}.$$

To show that $\mathrm{Cl}\,A$ is not path connected, we will show that there is no path from $(0,0)$ to $(\frac{1}{\pi},0) \in A$.

On the contrary, suppose that there is a path $\gamma : [0,1] \to \mathrm{Cl}\,A$ such that $\gamma(0) = (0,0)$ and $\gamma(1) = (\frac{1}{\pi},0)$. Let $\gamma(t) = (\gamma_1(t), \gamma_2(t))$. Since γ is continuous, γ_1 and γ_2 are continuous. Since γ_1 is continuous, by the intermediate value theorem there is $0 < t_1 < 1$ such that $\gamma_1(t) = \frac{2}{3\pi}$. Again, by the intermediate value theorem there is $0 < t_2 < t_1$ such that $\gamma_1(t_2) = \frac{2}{5\pi}$. In this way, we get a decreasing sequence (t_n) in $[0,1]$ such that $\gamma_1(t) = \frac{2}{(2n+1)\pi}$. Note that $\gamma_2(t) = \frac{1}{\sin \gamma_1(t)}$. Then $\gamma_2(t_n) = (-1)^n$. Since (t_n) is a decreasing and bounded below sequence, $t_n \to t$ in $[0,1]$. Since $(\gamma_2(t_n))$ is not convergent, γ_2 is not continuous at t, a contradiction.

Let $x,y \in \mathbb{R}^n$. Then the map $\gamma : [0,1] \to \mathbb{R}^n$ defined by $\gamma(t) = tx + (1-t)y$ is a line segment from x to y.

A set A in $\mathbb{R}^n$ is called convex if any two points of A can be joined by a line segment that is contained in A. We can observe that a convex set in path connected.

Example 7.3.9. An open ball $B(a,r)$ in $\mathbb{R}^n$ is path connected. Let $x,y \in B(a,r)$ and $t \in [0,1]$. Then

$$
\begin{aligned}
\|\gamma(t) - a\| &= tx + (1-t)y - a \\
&= \|t(x-a) + (1-t)(y-a)\| \\
&\leq t\|x-a\| + (1-t)\|y-a\| \\
&< tr + (1-t)r \\
&= r.
\end{aligned}
$$

Therefore the line segment $\gamma(t)$ is contained in $B(a,r)$.

Exercises

7.1. First of all, complete whatever is left for you as exercises.
7.2. Show that the normed spaces $(C[a,b], \|\cdot\|_\infty)$ and ℓ_2 are connected.
7.3. Is an open ball connected in a metric space? Can you think of some sufficient conditions for an open ball in a metric space be connected?
7.4. Use the intermediate value theorem to show that every real polynomial of odd degree has a real root.
7.5. If $f : \mathbb{S}^1 \to \mathbb{R}$ is a continuous map, then show that there is a point $x \in \mathbb{S}^1$ such that $f(x) = f(-x)$.
7.6. Let X be a connected metric space. Then a point $x \in X$ is called a cut point if $X \setminus \{x\}$ is disconnected.
    Suppose that x is a cut point of a connected metric space X. Show that if (A,B) is a disconnection of $X \setminus \{x\}$, then $A \cup \{x\}$ and $B \cup \{x\}$ are connected.

7.7. Show that if X is a connected metric space containing at least two distinct points, then $X = A \cup B$ for some nonempty connected sets A and B in X.

7.8. Show that the set A in $\mathbb{R}^2$ containing the points in $\mathbb{R}^2$ such that at least one coordinate is rational is connected.

7.9. Show that if X is a countable connected metric space, then every continuous map from X to the real line is constant.

7.10. Let $f : X \to Y$ be a surjective continuous map. Is the continuous image of a component in X a component in Y?

Bibliography

[1]   F. K. Dashiell Jr., Countable metric spaces without isolated points, Am. Math. Mon. 128 (2021), 265–267.
[2]   J. Diedonne, Foundations of Modern Analysis, Academic Press, New York, 1960.
[3]   N. J. Fine and G. E. Schweigert, On the group of homeomorphisms of an arc, Ann. Math. 62 (1955), 237–253.
[4]   R. A. Horn and C. R. Johnson, Matrix Analysis, Cambridge University Press, Delhi, 2013.
[5]   S. Katok, p-Adic Analysis Compared with Real, Student Mathematical Library, Vol. 37, American Mathematical Society, 2007.

Further reading

[6]   J. Borsık and J. Dobos, Functions whose composition with every metric is a metric, Math. Slovaca 31 (1981), 3–12.
[7]   J. Borsık and J. Dobos, On a product of metric spaces, Math. Slovaca 31 (1981), 193–205.
[8]   D. Burago, Y. Burago and S. Ivanov, A Course in Metric Geometry, GTM, Vol. 33, American Mathematical Society, Providence, Rhode Island, 2001.
[9]   F. Cagliari, B. D. Fabio and C. Landi, The natural pseudo-distance as a quotient pseudo-metric, and applications, Forum Math. 27 (2015), 1729–1742.
[10] P. Corazza, Introduction to metric-preserving functions, Am. Math. Mon. 106(4) (1999), 309–323.
[11] M. Gromov, Metric Structures for Riemannian and Non-Riemannian Spaces, Birkhäuser, Boston, MA, 2007.
[12] R. Hemasinha and J. R. Weaver, What are the linear isometries of 1_n^p?, Int. J. Math. Educ. Sci. Technol. 22(6) (1991), 945–951.
[13] J. L. Kelley, General Topology, Springer-Verlag, New York, 1975.
[14] S. S. Kim, A characterization of the set of points of continuity of a real function, Am. Math. Mon. 106(3) (2018), 258–259.
[15] M. Kwon, Dimension, linear functionals, and norms in a vector space, Am. Math. Mon. 117(8) (2010), 738–740.
[16] H. Maehara, Euclidean embeddings of finite metric spaces, Discrete Math. 313 (2013), 2848–2856.
[17] J. R. Munkres, Topology, Prentice Hall of India Pvt Ltd., 2000.
[18] I. Rosenholtz, Another proof that any compact metric space is the continuous image of the cantor set, Am. Math. Mon. 83(8) (1976), 646–647.
[19] I. J. Schoenberg, Remarks to Maurice Frechet's article "Sur La Definition Axiomatique D'Une Classe D'Espace Distances Vectoriellement Applicable Sur L'Espace De Hilbert", Ann. Math. 36(3) (1935), 724–732.
[20] A. J. Sieradski, An Introduction to Topology and Homotopy, PWS-KENT Publishing Company, Boston, 1992.
[21] W. Sierpinski, Sur une propriété topologique des ensenmbles dénombrables denses en soi, Fundam. Math. 1 (1920), 11–16.
[22] W. Sierpinski, Sur les ensembles connexes et non connexes, Fundam. Math. 2 (1921), 81–95.
[23] T. K. Sreenivasan, Some properties of distance functions, J. Indian Math. Soc. (N. S.) 11 (1947), 38–43.
[24] W. A. Sutherland, Introduction to Metric and Topological Spaces, Oxford University Press, New York, 2009.
[25] F. Terpe, Some properties of metric preserving functions, Proc. Conf. Topology, Measure and Fractals, Math. Res., Vol. 66, Akademie-Verlag, Berlin, 1992, 214–217.
[26] A. A. Tuzhilin, Lectures on Hausdorff and Gromov–Hausdorff distance geometry, Lecture notes.
[27] W. A. Wilson, On certain types of continuous transformations of metric spaces, Am. J. Math. 57 (1935), 62–68.

https://doi.org/10.1515/9783111636085-008

Index

https://doi.org/10.1515/9783111636085-009

www.ingramcontent.com/pod-product-compliance
Lightning Source LLC
Chambersburg PA
CBHW061406210326
41598CB00035B/6118